Five Inventions in Electro-fishing in India

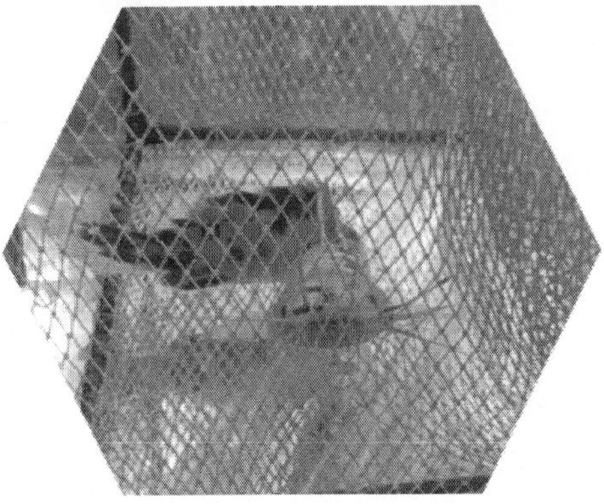

THE AUTHORS

Dr. K.P. Biswas, M.Sc., Ph.D., D.F.Sc. (Bombay), E.F. (West Germany), F.Z.S., F.A.B.S. (Kolkata) initiated the work on Electro-physiology and Electric Fishing, first in India, in 1961 at Odisha. He designed and developed electric seine, electric fish catcher, electric gill net, electric dip net and a new technique of harmless electric fishing, for which patent applications have been filed. He guided eleven M.Sc. students and one Ph.D. student of Calcutta University and four M.F.Sc. students of West Bengal University of Fishery and Animal Sciences for their thesis work on electrophysiology and fishing with electricity.

Miss Namrata Basu, M.Sc. from Calcutta University and a research scholar of Electrophysiological Laboratory, is actively associated with electrofisheries work for more than four years and was an active associate of the invention, *"A new technique of harmless electric fishing"*.

Five Inventions in Electro-fishing in India

K.P. Biswas

Namrata Basu

2021

Daya Publishing House®

A Division of

Astral International Pvt. Ltd.

New Delhi – 110 002

Cataloging in Publication Data--DK
Courtesy: D.K. Agencies (P) Ltd. <docinfo@dkagencies.com>
Biswas, K. P. (Kamakhya Pada), 1936- **author.**
Five inventions in electro-fishing in India / K.P. Biswas, Namrata Basu.
pages cm
Includes bibliographical references and index.
ISBN 9789390384853 (HB)

1. Electric fishing--India. 2. Electrophysiology. I. Basu, Namrata, author. II. Title.

LCC SH344.6.E4B57 2021 | DDC 639.2028 23

Published by : **Daya Publishing House®**
A Division of
Astral International Pvt. Ltd.
– ISO 9001:2015 Certified Company –
4736/23, Ansari Road, Darya Ganj,
New Delhi-110 002
Ph. 011-43549197, 23278134
E-mail: info@astralint.com
Website: www.astralint.com

Digitally Printed at : **Replika Press Pvt. Ltd.**

– Dedication –

The book is dedicated to the memory of
Late Mrs. Manju Biswas
for her inspiration as long as she was alive.

Acknowledgement

Authors gratefully acknowledge the moral support and help of Prof. N.A. Talwar and Miss Arpita Roy for preparing the soft copy of the manuscript of the book.

Dr. K.P. Biswas

Namrata Basu

Preface

Investigations on electro fisheries in India started as early as 1961 with the leadership of K.P. Biswas in Odisha. In the beginning he started working on the response and behavior of Indian fresh water fishes in underwater electric field with the object of catching fishes from deep and weed infested waters, where the conventional fishing gears fail to catch fish due to physical obstruction in body of water.

Laboratory tests were carried out with different varieties of Indian fresh water fishes to determine the threshold value of current intensities require to influence the fishes of different size group for different reactions, particularly the desirable galvanotaxis (attraction to the positive electrode). Simultaneously laboratory experiments were also done on microscopic animals (zooplanktons), fish food, available in natural water to find out the effect of electric exposure which the animal fish food (zooplankton) has to sustain during electric fishing in a water body containing both fish and their pray.

A challenging task of removing all predatory and unwanted fishes from nursery pond crop up for utilizing the pond for raising two crops fish seed without removing water in 1962, where the possibility of utilizing electric fishing for removing unwanted fishes were experimented in the field. Biswas successfully designed and developed an electrically charged net for removing unwanted aquatic animals including fishes from the pond.

These experiments were followed by designing and developing a large electric seine net for giving trials in large and deep water bodies to catch more fish. Forty six trials were given in 24 tanks of 1.2 to 3.0 ha having a water depth of 1.5 to 5.6 meter infested with marginal and submerged weeds. A control run without applying electric current was made for each electrical seining. Electric seine was found to have an increased catch over control seining both in numbers and weight. Bottom dwelling and burrowing fishes were found to be more sensitive and caught in large numbers in electric seine. Small size burrowers and bottom dweller have also been caught in large numbers in electric seine in addition to larger ones.

The work suffered a long gap due to the pre occupation of the author with the administrative work. The experimental work again resume from 2000 in the electro physiological laboratory of the author and the fishing gear laboratory of the West Bengal University of Animal and Fishery Sciences. Last twenty years study of electric fishing gear in flume tank resulted in designing and testing of five different fishing gears which can be used in 5 different conditions for more fish catch per unit effort. Patent applications have been filed with respect of all the 5 inventions described in chapter 1 to 5.

Dr. K.P. Biswas

Namrata Basu

Contents

Invention Teams

- **Team Members –** Dr. K.P. Biswas – Principal Investigator,
  Shri T.S. Murty – Associate

  Name of the invention – An Electrically Charged Net to Remove all Predators and Unwanted Fish from Fish Culture Ponds at Nursing and Rearing Stage

  Year – 1961–63

  Place of work – Odisha Govt. Fish Farm, Kaushlyagang, Dhouli, Near Bhubaneswar, Odisha, India

- **Team Members –** Dr. K.P. Biswas – Principal Investigator
  Prof. N.A. Talwar – Associate

  Name of the invention – Development of an Electric Fish Catcher

  Year – 2015–17

  Place of work – Electrophysiological Laboratory, 48, Boral Main Road, Kolkata – 700 084 Fishery Engineering Laboratory, WBUFAS, New Garia.

- **Team members –** Dr. K.P. Biswas – Principal Investigators
  Miss Namrata Basu – Associate
  Miss Arpita Roy – Associate
  Mr. Amitabho Roy – Associate
  Mrs. Nabanita Chakraborty – Associate

  Name of the invention – A New Technique of Harmless Electric Fishing

Year – 2019–20

Place of work – Electrophysiological Laboratory, 48, Boral Main Road, Kolkata – 700 084

- **Team members** – Dr. K.P. Biswas – Principal Investigator
  Mr. Subhasis Dolui – Associate

Name of the invention – Design of a New Three-walled Electric Gill Net to Attract and Catch Fresh Water Fish

Year – 2018–19

Place of work – Electrophysiological Laboratory, 48, Boral Main Road, Kolkata – 700 084

- **Team members** – Dr. K.P. Biswas – Principal Investigator
  Miss Shaik Kauser – Associate

Name of the invention – A Prototype Electric Dip Net

Year – 2018–19

Place of work – Electrophysiological Laboratory, 48, Boral Main Road, Kolkata – 700 084

The book, " *Five Inventions in Electric Fishing in India* " describes how a research endeavour could be nourished for such a long period to find out new ways and means to solve the problems of fish capture with the aid of electric stimulation against all odds and difficulties.

Neither sophisticated instruments and equipments for the work was available in the country, nor the trained and knowledgeable personnel to assist in the inventory work.

With the available materials and equipments in the country, how the new ways and means have been explored, one after another, to solve the problems.

Voices have been raised in many quarters about the illegal and improper use of electricity for fishing, which is not only harmful and causing injuries to fishes, but also has become the cause of fatality to the unauthorized applicators.

Proper use under the guidance of qualified and knowledgeable persons, the technique was proved to be very useful for fish capture and less harmful than other fishing gears.

The technique of harmless electric fishing described in this book will show a new way of using electric stimulation for mass scale fish capture.

The book will be useful for the students and will inspire the research workers.

Introduction

Electricity has the capacity for non-lethal stimulation to enable capture of fish. Any electric fishing system requires some minimum effective value of current density produced from the electrodes. The value of current density will vary with water conductivity, temperature and target fish. The current level in water establishes the perception, effective and danger zones in the field surrounding the anode.

As the current spreads into the water around any electrode, the current density will correspondingly drop off steadily as the distance from the electrode increases. Many references suggest that current is the variable most, directly related to electrical effects on fish. For this reason current density in the water is the best measure of electro-fishing effects.

All fish have approximately the same nervous system, which are similar to those of other vertebrate species. In a dorsal view, the nerve follows the myomeres out from the spinal cord and integrates the muscles. The frontal part of the brain appears to carry a negative charge, which may explain why fish are attracted to the positive electrode. The nerve angles and lengths appear to be important determinants of galvanotaxis (involuntary swimming to the positive electrode) as the orientation of fish to the electrical field determine how the fish will be affected.

Haskell (1957) and Vibert (1963) found that direct stimulation of muscle required considerably greater voltage than that was needed for neural reaction. Neural stimulation, therefore, was the most useful stimulus in electric fishing, because it occurred at lower electric intensities than muscle stimulation, but achieved similar results.

The objective of galvanotaxis is to interfere with the neurological pathway between the brain and muscles of fish. By blocking the internal signal and overriding it with a signal from the water, field currents redirect the neurological signal and muscular reaction.

Fish can swim undisturbed through an electrical barrier, if they are oriented in a particular position in the electric field. Likewise, once in the electrical field, the behavior of a fish varies with its orientation. The optimal reaction is involuntary swimming in the predictable direction (towards the anode). If the force of muscular contraction is too great, however, spinal injuries may occur.

Fish absorbs power as a function of body surface area and particularly length (Holzer 1931).

As current flow leaves the electrodes, passing through the water, it spreads out in all directions forming a field pattern. The conductivity of the water, which is determined by the concentration in the water of charge carriers (ions) influence the shape and extent of the electric field in water and thus affects the fields ability to induce capture-prone behavior in the fish.

Since electro-fishing is thought partly science and partly art, it may be possible to replicate results but impossible to achieve full comparability between studies, performed by different workers using different equipments in different conditions.

When the fish are aligned along the current lines, it will experience the greatest voltage potential. When aligned along the voltage lines it should experience the least voltage difference. However, they will experience some lateral voltage gradient across its body.

The fish response of avoidance, forced swimming, narcosis and tetany depend not only in the current type, but also, on the total duration and level of current density it experiences.

Direct current field will cause the fish to swim involuntary to the positive electrode, where they are stunned when they reach the strong field near the electrode.

Extensive strong directional swimming does not occur unless the DC (direct current) is pulsed. The reactions of fish to pulsed DC of increasing strength are similar to those reported for DC except that before directional swimming commenced, the pulsation set up a series of movements which appear to be galvanotropic reflexes of spinal origin. It was thought that locomotion might be a voluntary escape reaction at that time, but that galvanotropic reflex produced by pulsed DC inhibited the fish from turning away from the anode. The result is violent milling movement.

Pulsed DC is produced by the regulated interruption of the continuous flow of DC which yields a pulsed signal. The wave form of pulsed DC is like a hybrid between DC and AC. There is a good attraction power of pulsed DC, but this is due to the electrical effect on the fish's muscles (the muscles contracting at each pulse of electric current and thus accentuating the swimming motion) and not as in DC by electrical effects on the spinal nerves.

One of the most popular approaches at present to exit is "pulse trains" or "pulse energy packets" or a low energy wave forms rather than rectangular pulses, using less energy and reducing the rate of severity of injury of fish.

Low pulse rate (5-15 pulses per second) are similar to DC except they have greater effective range. Higher pulse rate (40 to 120 pulses per second) have greater range but fish will be stunned as they approach the electrodes.

Similar to the concentrations of most chemical substances and physical parameter affecting living organism, the electrical field intensity and duration of exposure are the primary factors affecting physiological stress and mortality in fish subjected to electric fishing currents. Beyond, it is not the field intensity itself but the magnitude of voltage differential generated across the fish (head to tail voltage) or specific nerves and tissues that causes mortalities in electro-fishing. Differential voltage is a function of both field intensity and orientation of fish relative to the lines of current.

Potential effect of fish injuries during electro-fishing can be - (a) cardiac or respiratory failure, (b) spinal or oth er related injuries and (c) stress and fatigue. With vertical dark bands occur in the area of dorsal fin where the muscle mass is the greatest. Fish with poor health may respond less strongly to the electric field and is less able to withstand the stresses of tetany and apnea during narcosis, thus increasing the possibility of death.

Stress related effects may be short or long term with implication on fish health. Neuro-endrocrine changes from exposure to electric field, changes the blood chemistry and hematology by the way of hyperglycemia, hyperlactemia, changes in blood plasma cortisol levels, plasma chloride, plasma protein, haematocrit value, leucocrit value and haemoglobin concentration, within two hours of shock and take 8 hours to reach the base levels. Reduced swimming stamina, lethargy and cryptic behavior have been noticed following electric exposure.

Spinal injuries such as muscular convulsions (myolclonic jerks or seizures) induced by sudden change in field intensity or more specifically in voltage differential across the fish or affected tissues at or above a relatively low threshold in magnitude of change for twitch. Sudden changes occur when the current is switched on and off or pulsed, or when a fish leap frantically out and back into electrified water. Slowly rising current have little effect in fish.

The causative factors of stress and injuries of fish in an electric field may be, fish species, sizes, physiological conditions beside, current type, field intensity, duration of exposure, orientation of fish with respect to current lines, pulse frequency, wave form and pulse width. Individual variations are notable among fish, even though they are of same species and have similar lengths.

The development of today's electric fishing gear have been directed towards reducing harm to fish, saving time and money and reducing the possibility of serious injuries to fishery workers and increasing the catch per unit efforts. If the electrical combination (level of voltage, pulse width and pulse rate) is not adequate, fish would immediately escape to the side or towards the cathode.

Increasing the voltage without increasing the electrode size increases both danger zone and effective zone and requires a much greater increase in power, since both voltage and current are increased.

Electro-fishing (passing electricity through water to catch fish) requires sophisticated equipment and trained personal. In India, none of these facilities are available. While this method is commonly employed by scientists for survey and fisheries management, a crude and illegal form of electro-fishing is employed for catching fish from water bodies in some parts of India. In Kerala people use a crude version of this method to catch fish from streams, ponds and flooded paddy fields. They generally use only the crudest safety mechanisms and even illegally divert electricity from over head electric lines. This occasionally results in accidental electrocution and death.

Eight cases of accidental death occurred in Kerala during electro-fishing by adapting the crude and illegal method without having proper training and equipment between January 2005 to November 2013.

Local fishermen of Loktak Lake, Manipur use crude version of passing electricity through water to kill fish in lakes and streams. This crude version of the method to kill fish in lakes and streams are being performed in many regions of India. Often diesel run generator or illegal use of overhead electric lines of the power grid is used to kill and catch fish. In Loktak Lake, fishermen were using a car battery to power electro-fishing. To make the shock fatal to the fish they were converting DC to AC power by using a power inverter. They would kill fish by inserting metallic rod and electric wire into small pools on the banks of the lake. The dead fish float up to the surface and would be picked up by hand to move on to repeat the process. They use very little safety measure and several accidental fatalities have occurred due to this practice. Beyond the risk of human injury and death, is the effect of such fishing practices on fish assemblages in fresh water systems.

Electro-fishing can yield much higher capture rate, compared to the conventional fishing methods, such as nets and fish traps, but there need to be regulated use of this fishing practice, particularly by the trained and knowledgeable persons to avoid accidents and damage that it cause to fish population.

1

Invention of Electrically Charged Net for Total Eradication of Predators and Weed Fishes from Fish Culture Pond

Conceived Design and Trial – Dr. K.P. Biswas

Associate in the Work – Mr. T.S. Murty

Background of Invention

In early sixties, fish seeds of Indian major carps (*Labeo rohita, Catla catla, Cirrhinus mrigala* and *Labeo calbasu*) were collected from the flooded rivers during monsoon, as the technique of hypophysation and induced breeding of major carps were not known. Since minor carps, minnows and some species of piscivorous fish also breed at the same time with that of major carps, the spawn (fish seed) available in the flooded rivers, were a mixture of major portion of unwanted species and a very small amount of major carps, depending on the population of major carp species in the river, which was in a vary dwindling state due to illegal fishing of brood stock from the river pools during dry months.

In the shooting net (used for fish seed collection) from the rivers during flooded season, a mixture of major carp spawn and spawn of unwanted species was collected. There was no technique available to identify and segregate major carps from spawn of unwanted species as they were too small (5 to 7 mm in length). As a result, manured and prepared nursery ponds stocked with a riverine spawn, after two weeks of feeding, if found with less than 20 per cent major carp fries (at fries stage of 25 -35mm size major carp seed would be distinctly identified), the entire stock had to be destroyed and the pond had to be again prepared for receiving new batch of riverine spawn.

But the problem remains for removing the unwanted species from the nursery pond. There were two alternatives to tackle the problem at that time, either to dry the pond by pumping out water, which normally took few days for complete drying of the pond bed and again to fill up the dried pond with water for manuring to receive a new batch of riverine spawn. Water was scarce prior to monsoon and there was a risk of entry of unwanted species if the pond is filled with the water from other tank. If drying and refilling of water is not feasible, the other way is to apply fish poison, like mahua oil cake in the pond water to kill the entire pond biota. In that case refilling of the pond may not be necessary but has to wait for 21 days to dissipate the toxicity of pond water treated with fish poison.

For these difficulties, it was not possible raise 2 – 3 crops of major carp fry from perennial nursery pond as riverine spawn collection from flood water of rivers lasted for 2 to 3 months in 2 to3 spells.

So the need was to develop a technique to kill the unwanted stock of nursery pond within a short period without poisoning or removing water for making the ponds ready within 2-3 days for instant manuring and to stock a fresh batch of riverine spawn to raise them to fry stage within 15 days, so as to obtain 2-3 crops in the monsoon season.

In Odisha, the State Fishery Department were given responsibility to supply quality major carp fry (fish seed) to the pisciculturist. As such every year there was a good amount of pressure demanding supply of quality fish seed to the fish farmers out of the riverine spawn collection. To mitigate this problem there were pressure on the fish seed producing centers for production of more quality fish seed during a short monsoon period.

In this context, complete removal of unwanted stock from perennial nursery pond within a short period, without removing or poisoning the water, the design and development of the electrically charged net was conceived by Dr. K.P. Biswas in 1962. The method was tested in fourteen nursery ponds in 1962 with great success.

Invention

Fishing with electricity is one of the most common methods used in fresh water fishery research and management, namely sampling, to know the status of fish population. The basic principle involves, creating an electric field in the water to immobilize fish. When fish are stunned, they often float near the surface of water and can be removed from electrical field. The function of an electro-fishing system is to produce an adequate electrical stimulus in fish body near the electrodes to permit the easy capture by netting. In an alternating current (AC) system, the electric stimulus only immobilizes fish (electronarcosis or electrotetanous) requiring a back-up net to collect fish from considerable depth or distances from with electrodes.

The primary function of an electro-fishing system is to establish an electrical field in the water near fish. A portion of the current (depending on the ratio of fish and water conductivity) will pass through the fish and if of sufficient magnitude will elicit the desirable response. Important point is that the current is a variable of concern. Electrodes, voltages, spacing of electrodes size and shape are the important only to the extent that they influence the current.

In order to remove predatory and other bottom dwelling fishes including burrowers from the rearing ponds before stocking cultivable fish seed, the use of derris root powder (rotenone) are the common practice among fish culturist but in India, scarcity of derris root powder necessitate the adoption of mahua oil cake and other pesticide for clearing of nursery pond. Each of this method has its own limitation and sphere of usefulness, especially of residual toxicity.

A vertical field was created and used for the field trials in 14 perennial ponds, meant to be use for rearing Indian major carp seed after completely removing the unwanted predators and competitors.

Alternating current (AC) of 50Hz was used in these trials.

Before conducting the field trials laboratory experiments have been carried out in an insulated glass tank of 70cm X 30cm X 65cm dimension. Two 6 mm solid

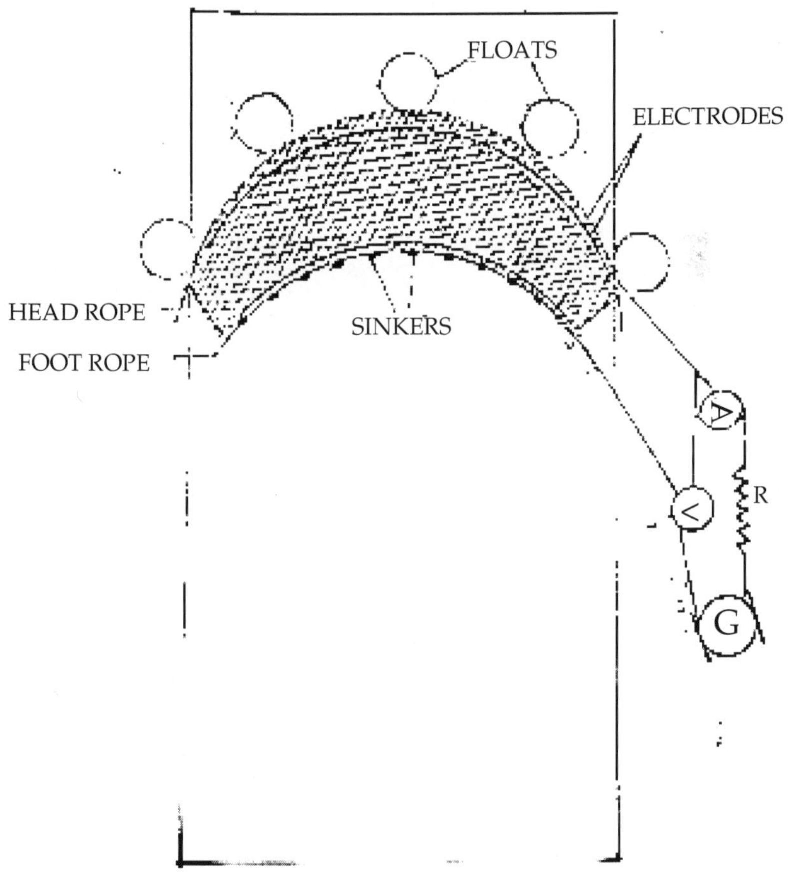

Schematic Diagram of Operation of Electric Net Showing Power Supply.

aluminum rod vertically separated at a distance of 48cm served as electrodes of the heterogeneous field. Alternating current 10 - 225V were applied to the electrodes and the field strength was measured by a probe of silver electrode (1sq. cm separated at a distance of 1cm from each other). The probe measured the voltage gradient and current density between the electrodes with the increasing applied voltages of 10, 20, 50, 100, 105, 135, 155, 175, 195 and 225V. The field patterns away from the electrodes by 7, 14, 21, 28, 35, 42cm at both sides were measured for current densities and voltage gradient.

The net consisted of 20m long and 6m wide fine meshed (5mm stretched mesh, so that minnows and small predatory fish cannot escape) cotton net with interven 9mm dia sisal head rope at the top and 11mm dia sisal foot rope at the bottom. The head rope was provided with 10cm dia, synthetic floats attached, 1.2m apart from each others. Lead sinkers (barrel shaped weighing 250gm each) were attached to the foot rope at the interval of 1m.

Basing on the data of the laboratory experiments the electrical net was designed for field trials in earthen nursery ponds.

Seven strands of flexible bare copper wire of 16m length, each served as electrode, which was interwoven along with the foot rope of the entire net leaving a gap of 0.5m in both sides of the net. The other electrode (bare flexible copper wire) was fitted along the body of the entire net length of the net, 0.3m below the head rope and leaving a gap of 0.5m in both sides of the net, like the other electrode attached to the foot rope. The electrodes both at the top and the bottom remain within the water during seining operation. In fact, the electrode interwoven with foot rope was ploughing through the bottom mud and silt, while the net was dragged from one end of pond to other. The vertical distance between the electrodes in the water depended on the depth water in the pond (0.85 – 1.3m) and lines of current conduction will flow vertically through the entire column of water between the two bare copper wire electrodes.

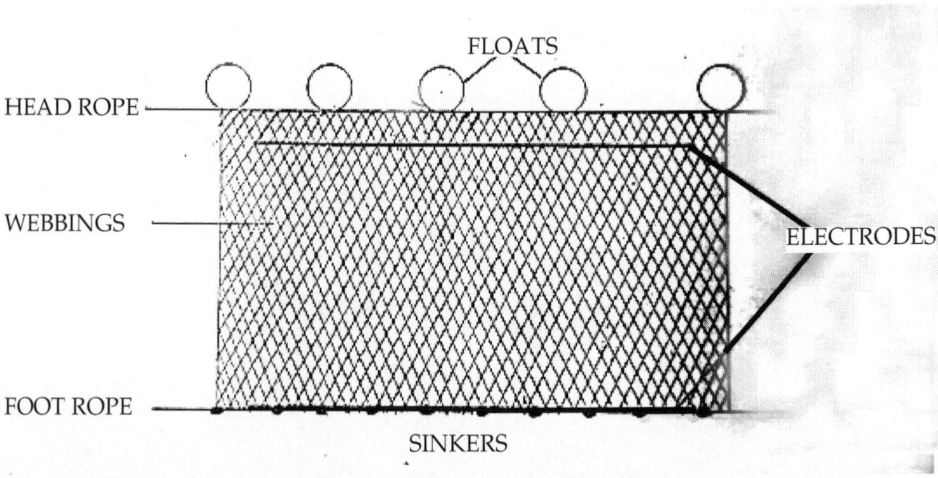

Schematic Diagram of Electric Net Showing the Position of Electrodes.

While connecting the electrodes with AC power supply one precaution was taken so that, the phase was connected to the electrode near the head rope and neutral line was connected to the foot rope electrode. With this configuration, a moving vertical electric field between the electrodes would be created along the entire water column of the pond and to the entire water area of the pond, when the net moved from one side of the pond to the other.

The power supply equipments, mainly consisted of one 3Kw AC generator, mounted on a trolly, driven by 5HP water cooled kirloskar engine to deliver 230 volts at 1500 rpm. The current form the generator output passed through a switch board consisting of voltmeter, ammeter to measure the applied potential difference in volts and the amount of current drained in amperes during the dragging operation, was connected to the electrodes through an insulated conducting cable. The phase of the output terminal was connected to the head rope electrode, while the neutral line was attached to the electrode of the foot rope, so as to avoid short circuit of current flow to the earth.

The mode of operation of the gear did not lead to any complexity. Four persons pulled the net, two on each side of the net without wading.

To determine the effectiveness of electrically charged net with that of conventional one (without application of electric field) the same net was operated in the pond first, without allowing the electric current to flow through the electrodes at each of the experimental pond. The catch of conventional net was analyzed for species size, number and weight individually, except in case of small fishes and minnows, where the number and weight of fishes were taken. The entire catch of the conventional net, after keeping the measurements, were released back alive in the pond to maintain fish population same as before.

The drag net was then electrically charged by connecting the power line to the electrodes fitted to the net, only after putting the net in the pond water. The charged net, with electric field around the electrodes was then dragged from one end of the pond to the other. The voltage between the electrodes dropped to 100 to 140 volts in 13 ponds and to 50 volts in one pond while dragging the net from one end of the pond to the other, due to heavy drainage of current between 11 to 15 amperes, which was due to varied pond soil conductivity. The current flow to the net was switched off after completion of dragging operation. Fishes moving in the pond water column were influenced by the vertical electric field of the moving electrically charged net and found to float near the head rope electrode in a state of narcosis. The bottom dwelling and burrowing fish coming within the effective range of foot rope electrode were lifted in the water column between the electrodes and found suspended in the mid-field between the electrodes in hypnotic condition. The collected fish in the charged net was analyzed for species, numbers, size and weight in the same manner as was done for catch of conventional net.

The electrical resistance of the pond water was measured with a probe and was expressed as ohms per cubic centimeter.

Each of the experimental pond was netted six times with electric net and the catch compared with conventional net.

The percentage of excess catch in electric net over conventional net was calculated by number of fish catch in electric net minus number of fish collected in conventional net (difference) divided by the number of fish caught in conventional net multiplied by 100.

The pond was finally drained to dryness to check any left-over fish that escaped the electric net.

Significant Findings

The effective zone of the heterogeneous electric field created by bare copper wire electrodes in the pond water vertically, was found to be the widest on the electrode-line and gradually turned out to be narrower as the distances from the electrodes increased. Testing with live fishes in the laboratory set up, *Labeo rohita* (Indian Major Carp), *Channa punctatus* (murrel) and *Cirrhinus mrigala* (bottom dwelling Indian major carp) a current density of 125μA/cm^2 produced by 45 volts influenced them for narcosis.

The percentage of excess catch in electrically charged net over the catch of ordinary net without applying electric current in 14 experimental ponds ranged from 4 to 310 percent in all the ponds except one pond (pond no 5). The excess catch of 50 per cent or more in electric net were noticed with respect to bottom dwelling and off-bottom fishes hiding under cover. They were *Cirrhinus mrigala, Labeo calbasu, Cyprinus carpio, Labeo gonius, Labeo bata, Cirrhinus reba, Channa spp., Glassogobius giuris, Mystus spp., Rhyncobdella aculeata* and *Clarias batrachus*.

The dropping of voltage at electrodes from 230 volts (generator output) to 50 to 140 volts at electrodes while applying load during electric netting operation was different for different ponds affecting the amount of current drained during electric netting. Fifteen to thirty five minutes were required for each electric netting, during which period 11 to 15 amperes were drained lowering the voltage from 230 volts to 100 to 140 volts, due to excess load on the generator.

The significant increase of catch in electric net over conventional net was noticed in case of *C. mrigala* (400 per cent), *Channa spp.* (120 per cent), *C. batrachus* (100 per cent) in pond with water resistance of 18000 ohms per cubic centimeter. In lower water resistance of 16000 ohms/cm^3, *L. calbasu* (100 per cent), *C. reba* (335 per cent), *Channa spp.* (2100 per cent), *G. giuris* (180 per cent) and *Mystus spp.* (282 per cent) were caught more in numbers in electric net. In the next lower water resistance (15000ohms/cm^3) species like *C. mrigala* (750 per cent), *L. bata* (100 per cent), *C. reba* (100 per cent), *Channa spp.* (500 per cent) and *Mystus spp.* (156 per cent) were captured in excess numbers in electric net, than the ordinary non-electric net. In still lower range of water resistance of 10700 to 6000 ohms/cm^3, *C. mrigala* (137 per cent in 6000 ohms/cm^3), *L. calbasu* (284 per cent in 9300 ohms/cm^3), *L. bata* (200 per cent in 6000 ohms/cm^3), *C. reba* (100 per cent in 9300 ohms/cm^3), *Puntius sarana* (100 per cent in 6000 ohms/cm^3), *G. guiris* (117 per cent in 6000 ohms/cm^3), *Mystus spp.* (100 per cent and 200 per cent in waters of 6000 ohms/cm^3 and 9000 ohms/cm^3 respectively), *P. ticto* (200 per cent in 10700 ohms/cm^3) were caught in excess numbers in electric net when compared with the catch of non-electric ordinary net.

With respect to availability of different species of fish of different size groups in experimental net, the percentage of excess catch of 100 per cent and more was found in *L. rohita* of 25-30cm length. In the next lower size range of 20-25cm, *C. mrigala* (370 per cent), *L. calbasu* (811 per cent), *Channa* spp. (228 per cent) and *C. batrachus* (100 per cent) dominated the catch of electric net. *L. calbasu* (173 per cent) and *Channa* spp. (420 per cent) predominated in the catch of electric net at 15-20cm size group. In the 10-15cm length group, *L. calbasu* (194 per cent), *L. bata* (100 per cent), *Channa* spp. (242 per cent) and *G. giuris* (500 per cent) were found to be more in the catch of electric net. Both *G. giuris* and *Mystus* spp. were 200 per cent to 240 per cent and 376 per cent to 177 per cent more in the size range 5-10cm and 1-5cm respectively. An excess catch of 650 per cent *C. mrigala* in electric net was recorded in the length range of 5-10cm.

Repeated electric netting up to six times in each pond and the percentage of removal of fish after each electric netting revealed 56.3 to 100 per cent capture in the first netting in 14 ponds, 6.7 to 33.3 per cent in the second netting in 13 ponds, 3.7 to 27.5 per cent in third netting in 11 ponds, 0.6 to 22.1 per cent in the 4th netting in 11 ponds, 0.6 to 7.2 per cent in 5th netting in 9 ponds and 1.1 to 6.0 per cent in 6th netting in 5 ponds, totalling a capture of 97.3 to 100 per cent in all the 14 ponds.

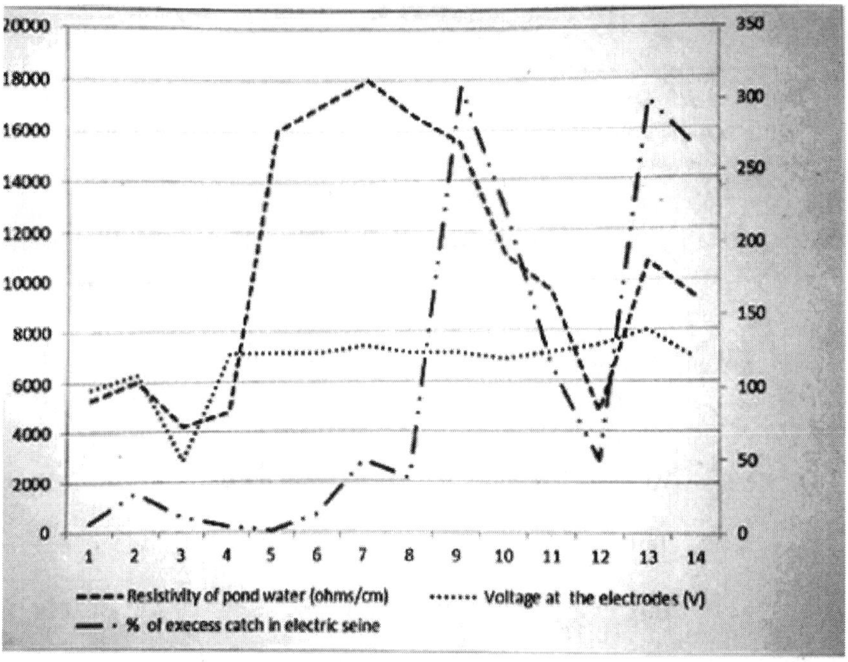

Characteristics of Electric Netting in 14 Experimental Ponds Showing the Percentage of Excess Catch.

After six electrical netting in each of the 14 ponds, all the ponds were drained to dryness to check the left over fish if any. The removal of fish from each pond progressively declined at each of subsequent electric netting and was lowest in the sixth electric netting. The total capture of electric net after six netting amounts to 97.7 to 100 per cent. Only 0.1 to 2.3 per cent leftover fishes remained in 5 ponds out of 14 ponds netted with electric net.

Claims of Invention

The electric field strength and pattern of the electric field, similar to that developed by electric net used in pond trials, created in the laboratory, indicated that the most effective vertical zone to influence fish was just along the line in between the electrodes, where the current density of 440 – 450 $\mu A/cm^2$ was maintained depending on the pond water and bottom soil conductivity. This value of current density progressively decreased to $205\mu A/cm^2$ as the distance away from the electrodes increased to 42cm from the electrodes. Experimenting with live fishes confirmed the effectiveness of the field strength of the vertical field.

The electrodes fixed in the electric net remained in close proximity (0.55 - 1.00 m) in the pond water able to create a strong vertical electric field, which was just sufficient enough to cause oscillotaxis and electronarcosis in fish and swept in the net behind, while dragging the net from one side of the pond to the other side.

Of the fourteen experimental ponds, an excess catch (all fishes combined) in electric net over the conventional one was observed in 13 ponds. The excess catch was highest (310 per cent and 300 per cent) in pond no 9 and 13 in water resistance of 15500 and 10700 ohms per cubic centimeter. The lowest excess catch of 4 per cent can be attributed to the higher conductivity of water in pond no. 4 (4800 ohms/cm^3), than the fish body enabling to pass a smaller portion of current through fish to elicit electronarcosis (Denzer, 1956).

Though the conventional net can encircle a large portion of fish population of the pond, the percentage of escape through the top, sides and bottom of the net remain large. The fish generally escape by jumping over the float line and diving through silt and weeds at the bottom during dragging and hauling the net. In the electric net the fish population of the pond being influenced by the moving vertical electric field created in the water column between the electrode exhibited oscillotaxis followed by narcosis, did not get any chance for jumping out and escaping through the sides and under the foot rope when they enter the effective zone, reducing the percentage of escape to a minimum. This was evidenced, when the bottom dwelling and off bottom fishes, like, *C. mrigala, L. calbasu, Channa spp, Glassogobius giuris, Mystus spp* and burrowers like *Rhyncobdella aculeata* were caught more in numbers in electric net than conventional net.

The electric net used in the experiments was of cotton webbing into which electrodes were fitted. The purpose of using net webbings in electric net was to collect the influenced fish in the net behind, instead of trying to collect them by a dip net, a possibility of escaping fish in deep waters. The chances of fish escape will further increase when the fish after receiving a mild electric shock in the peripheral zone will be frightened and move deeper and outside the electric field, had there been no net behind.

To compare the efficiency of electric net, each time before the operation of the net with electric charge, it was operated without application of electric current. Biswas (1970) used the same technique in large meshed (5 cm in stretched condition) fishing net to find out the possibility of catching more fish in electric net per unit of effort. He found that the electric net had an increased catch over the control net both in numbers and weight. Benthic fishes, like, *C. mrigala, Glassogobius giuris, Channa spp.* and *Mystus spp.* were found to be more sensitive to electric net and an increased catch of smaller ones have been observed in electric net in addition to larger ones.

Godfrey (1956) reported the capture of eels, sculpin and burbot by repeated electro-fishing, which otherwise were usually difficult to capture with nets.

The present experiments aimed for complete removal of predatory and unwanted fish species from a battery of 14 nursery ponds without application of fish poisons in order to make the ponds readily available for rearing cultivable fish seed without heavy mortality due to predation.

Pratt (1951), while measuring the efficiency of alternate and direct current fish shocker, advocated the use of direct current (DC) fish shocker for collection of fishes, as DC was found to be less harmful to fish than AC and the shocked fish could be salvaged and returned to the water alive.

But the present work who designed to kill and remove all the unwanted fishes, which if allowed in the pond, even in few numbers, will cause large scale mortality of cultivable fish seed.

By repeated electric netting for six times in each pond, a very small fraction of total number of fish (0.1 to 2.3 per cent) remained in 5 out of 14 ponds (ascertained after draining of ponds to dryness) indicate that the nursery ponds could be made free of predators and competitors with repeated netting by electric net.

Bottom dwelling and predatory fishes like C. *mrigala, L. calbasu, Channa spp., G. giuris;, R. aculeate, C. bratrachus* and *Mystus spp.* were more sensitive to electric net and were caught more in numbers in electric net. This type of specific sensitivity was also observed by Smith and Elson (1950) with respect to eels towards the electric fishing gear.

Funk (1947) reported that the electric net collected large numbers of small sized long gear sun fish and green sunfish in the upper gasconde river, though large sized fishes escaped from the electric fields easily. But in present electric net, in addition to larger fish groups, small sized *C. mrigala, G. giuris* and *Mystus spp.* were caught in more numbers, suggesting the effectiveness of the use of electric field in fish catch, the majority of which otherwise could have escaped through the net without electric field around it as in case of conventional net.

Electric net was found to function most effectively in waters having a resistance of 9300 to 18000 ohms per cubic centimeter, since a higher voltage (120-140 V) at the electrodes could be maintained during dragging operation. In the lower resistive water (4200-6000 ohms per cubic centimeter) the open circuit voltage sometimes dropped down as low as 50 volts.

A combination of 120-140 volts at the electrodes in water resistance of 9300-18000 ohms per cubic centimeter was found to be most beneficial in the electric net designed and experimented in 14 earthen nursery ponds.

After a thorough and repeated netting (six times) with electric net in each of the 14 ponds, the total capture were found to be close estimation of total number of fish that were present in the pond. In all cases, after the final draining of pond water to dryness, only a small fraction (0.1-0.2 per cent in 28 per cent ponds and 2.3 per cent in one pond out of 14 ponds) of total number of fish remained in the ponds did not greatly question the success of new electric drag net.

Justification for the Claims

Comparative study of some of the prior arts related to an electrically charged net for removing predators and unwanted fish and aquatic animals from farm ponds

Pub. No. IN 1290/MUM/2012 – Mosquito attraction, collection and killing mosquito by electric charge

Conventionally mosquitoes are repelled by means of chemical evaporation and a smoke by burning of chemicals. The mechanical filters in the form of net of cotton and synthetic material are used to protect the human beings from mosquito bites. In the present invention the mosquitoes are attracted toward the machine by means of ultra frequency sound generated by means of integrated circuit with output device as micro speaker, odor and the dark color for which the mosquitoes are having affinity. All mosquitoes are collected in a chamber and killed by electrically charged net, sinking into the water, contact with hot plate, sinking into the chemicals or by crushing.

The device, electrically charged net is operated in ponds and tanks, particularly nursery and rearing ponds, meant for complete removal of unwanted predatory fish and animals from the pond to be used for rearing baby fish of commercially important varieties and is entirely different from earlier patent no. IN 1290/MUM/2012 which was for collecting mosquitoes from atmosphere and killing them.

Pub. No. CN 109349186 – Using method of floating net cage culture light trap

A buoyancy system composed of a floating plate and floating ball is arranged in a net cage. A rope is used to drag floating ball to fix the floating plate to enable a photovoltaic panel to face the sun direction of high noon and electric energy generated by the photovoltaic panel is stored in a storage battery. At nightfall an automatic controller starts control the light trap and a high voltage power grid are powered on. Winged insects are attracted and gathered around the light trap lose flying ability, when touching the high voltage power grid, fall into the water when they are eaten by chub and big head carp. The light trap can attract and gather plankton in the water areas, so as to increase the supply of plankton in the net cage. This floating net cage culture light trap claimed to be simple in design, convenient to use and good in effect and has popularization value.

The electrically charged net for which patent application was made is a device for narcotizing and killing unwanted predatory fish and other aquatic animals, from nursery and rearing pond prior to culture of economical baby fish, without drying the pond or using fish poisons. In both the cases it is time taking (about 20-25 days) to remove the residual toxicity of poison and expensive for refilling the pond with clean and fresh water. As such, the electric net is entirely different from floating net cage culture light trap.

Pub. No. CN 208523525 – Deep sea aquaculture net cage with solar electric system

A deep sea breeding cage with solar power generation system comprising of a solar panel, a grid and an aerator, characterized in that, the grid. The outer side at the intermediate position is provided with a netting and the top end of the netting is provided with a cover plate and the top end of the netting frame above the netting is provided with a bracket. A solar panel is mounted on the top end of the bracket. A battery is also mounted on the bracket below the solar panel and an output end of the solar panel is mounted on. The photovoltaic controller is electrically connected to the input end of the battery. The floating body is installed at both ends of the bottom of the grid and the floating body is closed to the mesh frame at the end of the mesh. The bottom end of the netting frame is mounted with an anchor body through traction chain and aerator is mounted under the netting via a traction rope and the netting is attached uniformly provided with a mesh. The wire is evenly wound and is evenly wrapped with cotton yarn.

The deep sea culture cage with a solar power generation system according to the claim characterized in that the longitudinal section of the solar panel has upwardly convex arc shape; the grid is uniformly installed with illuminison lamp at both end; an inflation hole is provided at an intermediate position of the bottom end of the floating body and that the outer side of the wire is provided with an oxidation resistant layer.

The electrically charged net n, a device to eradicate and remove all the predatory and unwanted fish and aquatic is animals from fresh water pond so as to make the pond ready for nursery for commercially important baby fish, instead of applying poison and drying the pond which is time consuming and expensive.

Pub. No. NL 8603257 – Drag net for fishing vessel- has electric current to disturb fish on sea bottom

The assembly is trailed behind the fishing vessel by a cable. The cable is offset from the ship's side by a boom. The lower front edge of the net is held open by a horizontal beam. Each end of the beam is held in a glide-shoe. Strung back through the net at right angles to the beam are electrodes. The electrodes deliver electric pulse current which disturb bottom feeding fish causing them to swim upwards where they will be trapped in the net. The pulse current is generated by a circuit housed inside one of the glide–shoe. The generator is powered by a rechargeable battery also housed in the glide-shoe.

The device under Pub. No. 8603257 was used in the sea, possibly in a beam trawler and the purpose was only to disturb the fish from the bottom to rise up and enter into the net. The device, therefore, is not at all comparable with the present electric drag net as it was used in fresh water ponds manually from the pond embankment to make the pond free from unwanted fish and predators for rearing baby fish instantly. The device was powered by a portable 3Kw AC generator supplying 230 volts.

Problem Solved by the Invention

Prior to adopting induced breeding technique by injecting pituitary hormone to Indian major carp (catla, rohu, mrigal, calbasu) to spawn in confined water, earlier to 1961, seeds (spawn) of Indian major carp wee collected from rivers during rainy season, where major carps used to breed in flooded rivers. The collection of fish seed from river lasted for 2-3 month in spurts. Other predatory and unwanted fish also breed at the same time with major carps. So the fish seed collected from river were mixture of major carp and other unwanted species. At the spawn stage (5-7mm length) it was not possible to identify major carp seed from other unwanted species unless they grow to 25-35mm size after 15-20 days rearing. If the reared stock was found to dominant unwanted species (more than 80 per cent), the stock has to be destroyed and start rearing of spawn for second time, for which the stock of unwanted species has to be completely removed. This was generally done with fish poison like 2-4-D or mahua oil cake. In either case the rearing pond has to be left for three weeks for detoxification and manuring to receive fresh batch of riverine spawn. As a result it was difficult to obtain 2/3 crops of fish fry from the rearing pond in a season of 2-3 months. It was therefore, been necessary to find out an alternative method to remove the stock of unwanted species from the rearing ponds within 2-3 days without applying poison or draining of pond water.

The invention of electrically charged net solved the problem of eradication of predators and weed fishes from rearing pond within 6-8 hours by netting with electrically charged net consecutively for six times to make the pond free from unwanted fishes, without draining water from the pond, to use for stocking of fresh batch of fish seed for second or third time by instant manuring.

Solutions Provided by the Invention

By the electrically charged net, the wastage of time and labor for removing predatory and unwanted fishes and aquatic animals have been saved and two to three crops of fish seed in a season of three months could have been possible.

Advantages

1. Saved time and labor for removal of unwanted fishes and predatory animals from the pond, keeping the water intact, for rearing fresh batch of fish seed of economic variety.
2. Higher percentage of fish can be caught from fish ponds (higher catch per unit effort) by electric fish net on the same principle of electrically charged net. The percentage of fish escape from conventional drag net by jumping

above the float line and diving beneath the bottom line can be reduced to a minimum, as fishes under the influence of a vertical electric field of a electrically charged net remain in a state of hypnotic condition as long as the current flow through the net. As a result the percentage of catch in a electric net become substantially high.

2

Development of an Electric Fish Catcher

Designed and Developed – Dr. K.P. Biswas

Associate in the Work – Prof. N.A. Talwar

Background of Invention

Basically, the function of an electro-fishing system is to produce sufficient electrical stimulus in fish near the electrodes to permit easy capture by netting. In AC system the electrical stimulus simply immobilize fish (electro narcosis or electro tetanus) requiring the net handler to collect fish from considerable depths or distances from the boat. Often fish was immobilized at depths beyond the reach of the dip netter and were not susceptible to capture at all. In DC or pulsed DC system, fish near the anode exhibited forced swimming (electrotaxis) towards the anode and hence more easily netted. It is well established that fish were much more susceptible to AC than to DC and hence the radius of action of an AC electro-fishing system operating at a particular voltage is much greater than a DC system at the same voltage. Pulsed DC is more effective than DC and to some extent combines the desirable forced swimming response of DC with the larger radius of action associated with AC.

The results regarding electro-fishing effects, indicated by Cuinat (1967) and of earlier work on stream shocking (Novotny and Priegel 1971) indicated that the primary function of electro-fishing system is to establish an electric current in the water near the fish. A portion of this current will pass through the fish and if of sufficient magnitude will elicit the desired response. The important point is that the current is the variable of concern. Electrode voltage, spacing, size *etc.* are important parameter only to the extent that they influence the current.

Based on this concept, for any electro-fishing system there exist some minimum value of current per electrode which will affect fish out to some specified distance from the electrode. The minimum current will depend upon electrode shape, nature of current, desired range of action, fish species, water temperature, water conductivity and many less obvious factors. Of these factors, only electrode shape and type of current are under control of designer. All other factors are external variables which must be compensated for by variations in the operation of electro-fishing system. Based upon the surveys, field tests and literature review carried during the study, the following specific problem areas could be identified;

Limitation Range

The distances at which fish are affected can be too small to produce useful sampling results. This is often caused by incorrect electrode arrangements and insufficient current and power.

Water Conductivity

Low water conductivity makes it difficult to attain sufficient currents to produce useful electro-fishing response. Lenon and Parker (1958) who found extreme conductivities in Appalachain mountain streams, attacked this problem by adding salt to the water to improve electro-fishing. Extremely high water conductivities call for currents too large to be supplied by portable equipment without special electrical control methods. For very high conductivity DC is ineffective. Other factors that are important are voltage and electrode size.

Water Clarity and Vegetation

Turbid water and excessive vegetation restrict visibility and reduce the value of the immobilizing capabilities of the AC electro-fishing. Pulsed DC and DC offer potential solutions to this problem.

Water Depth

Fish which are immobilized at depths exceeding 0.9 to 1.2 m are very difficult to capture by ordinary netting procedures. Pulsed DC and DC should overcome this problem.

Bottom Materials

High conductivity bottom materials tend to "short circuit" the current out of the water into the bottom material reducing the electro-fishing effectiveness and sometimes overloading the power source. Little can be done to remedy this situation except to use short electrodes, near the surface to avoid direct contact between electrodes and the bottom.

Water Temperature

The conductivity of the water and water temperatures are more or less a straight line relationship, that is, as the temperature of water increases, the conductivity of the water increases at an equivalent rate (Sigler, 1969). Theoretically success of electro-fishing should increase with temperature, but the workers in the field have noted differently. Smith and Elson (1950) believed that salmon parr exhibited the best response below 25°C and suckers at less than 20°C. Webster, Forney, Gibbs, Severus and van Woert (1955) had greater success in shocking brown trout at 7.8°C than at 16.7°C when both AC and DC were used. The response of brook trout to DC current was low in cold water, but increases with the temperature to 10°C. As the temperature increases further, the response decreases (Elson, 1942). Most salmonids are more easily captured by electro-fishing when the water temperatures were low 0 to 10°C (Vincent, 1971).

Fish Mortality

Mortalities caused by AC electro-fishing are higher than those caused by DC or pulsed DC (Taylor *et al.*, 1957) and the gross physical damage from AC can be severe (Hanck, 1949). Harmful effects from pulsed DC are usually a result of excessive exposure or intense electrical field (Pugh, 1962). It can be easily demonstrated that fish can be killed with electricity in their natural environment, as well as under laboratory conditions (Pratt, 1954, Godfrey, 1956).

With electro-fishing mortality is usually associated with regions of excessive high current density close to the electrodes. Potential solution involves electrode designs which avoid such regions of high current density by using larger electrodes and lower voltages. No injury to fish has been observed with DC or pulsed DC.

Fish Size

Individual variation is notable among fish even though they are of the same species and have similar lengths. The larger the individual of a species, the more sensitive it is to a given electric shock (Mc. Millan, 1928, McLain and Nielsen, 1953, Taylor *et al.*, 1957). Fish absorb power as a function of body surface area and particularly length (Holzer, 1931). Also, the greater resistance of smaller salmonids and possibly small fish of other species as well, further reduces their response to shock (Nakatani, 1954). All these factors contribute to capturing a greater proportion of large fish than were actually present in the population.

Fish Species

Some species are quite difficult to capture by electro-fishing. The most notable are northern pike and muskellunge. It appears likely that this is caused by the strong swimming ability and natural tendency to escape danger by rapid swimming coupled with high sensitivity to electric fields.

Large differences in response have been observed in various fish species when pulsed DC was used. Some selectivity in electro-fishing may be possible because of the differences. There may be optimal wave forms and pulse rates which will produce selective and efficient electro-fishing for various species.

Equipments and Operating Problems

Numerous problems associated with equipment limitations and lack of operating guidelines exist. Examples include inadequate lighting systems, insufficient power, lack of voltage control, lack of proper instrumentation, poor electrode design *etc.* the problems identified demonstrate the need for developing guidelines for electro-fishing systems capable of adjusting for variations in external conditions.

The Invention

Electro-fishing – catching fish with electricity started in different regions in different environment. Beginning with the use of storage cells and there after using induction coil to augment the voltage for catching fish, studies also took place in Germany on the orientation and movement of fish exposed to direct current (Mach, 1875; Herman, 1885). In England, Loeb and Maxwell (1896) demonstrated that the reactions were due to the stimulation of motor neurons and were forced not voluntary. Blasius and Schweitzer (1893), however discovered the phenomenon of "galvanonarcosis", a state in which the fish appears to sleep with a relaxed body when it is facing the anode. Contrastingly, a very strong stimulation was observed when the fish was facing the cathode.

The galvanotactive reaction of the fishes were utilized to develop the catching devices by creating an electric field in the water within the catching area. Fishes at their threshold current densities swim involuntarily to the positive electrode, where they are stunned when they reach the effective zone near the catcher electrode. This facilitates to catch the induced fishes unharmed for investigation or for commercial purposes (Lamarque, 1967).

In many countries, other than India, the devices are available in some compact unit like, Mofix – 250; Bafix – II in Germany, both for laboratory investigations and field use.

In the absence of such privilege in India, attempt has been made to develop a proto-type unit initially for laboratory studies, which can be extended to field use subsequently.

The electrical fish catching unit has been developed and tested to catch Indian fresh water fishes from the tank. In other countries mobile electric fish shockers have been developed mainly for fish population studies in rivers and streams.

The experimental unit had three main components namely; a) a holding tank, b) DC and pulsed DC power sources and c) the catcher electrode. For basic data on the threshold current intensities and behavior of Indian fresh water fishes, tests were carried out in homogenous electric field in standing and flowing waters in continuous direct current (DC) and pulsed direct current (PDC) of three frequencies.

The basic data thus obtained with regard to fish species, like, *Labeo rohita, Oreochromis mossambicus, Mystus tangra, Trichogaster fasciatus, Notopterus notopterus, Monopterus cuchia, Rhyncobdella aculeate* and crocodile fish have been used in designing and fabricating the catcher unit.

Initially some laboratory tests were done to establish the physiological status of the fishes to be used in the experiments to testify the suitability of electric catcher.

All experiments were conducted indoors in a glass tank of 1.2 m long, 0.45 m wide and 0.45 m deep. The tank was filled to a depth of 23 cm with water supplied by corporation. The cross – section profile of the tank was equipped with two 18 gauge thick aluminium plate electrodes positioned perpendicular to the longitudinal axis of the tank. Electricity (continuous and pulsed DC) was provided from the AC electric supply line through a step down transformer, rectified to DC and with an electronic pulser of low frequencies to produce pulsed DC.

The power source consisted of AC from supply lines of 240 V, which was stepped down through a 3.5 kVA transformer with eight tapings in the secondary from where any one of the following eight voltages could be obtained. Four diodes of 450 V were connected in series to any of the eight output voltages to rectify and smoothen by two capacitors of 330 µF to obtain DC of 50, 105, 130, 155, 175, 200, 220 and 240V respectively. In the external circuit two variable resistances of 1000 and 500 ohms were provided before supplying to the electrodes. The maximum potential supplied was about 240 V, the minimum unloaded voltage was about 50 V.

Two types of electrodes were used. For the negative (-ve) pole two rectangular aluminium plate (29 X 29cm) and for the positive (+ve) pole a helix coil of thick (3mm) copper wire fitted to a wooden handle was used. The handle was provided with a push button switch through which the positive pole was connected to the helix coil.

For test experiments two aluminum plates were used, one for the positive and the other for the negative electrode to create a homogeneous electric field. But in case of catcher electrode (helix coil) which was connected to the positive pole, either single or two aluminium negative electrodes were tried either singly or doubly in the vertical and horizontal plane of the experimental tank. The schematic diagram of the power source used in test experiments and the catcher electrode is given below.

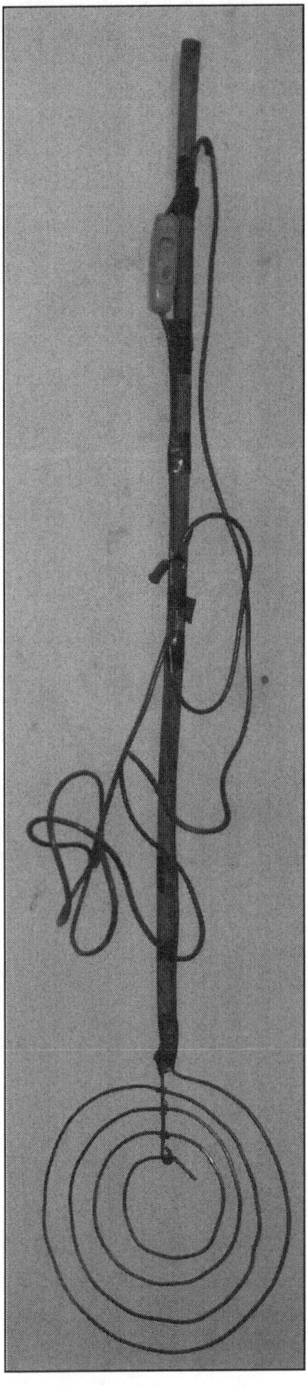

Helix Coil Positive Electrode Used as Catcher.

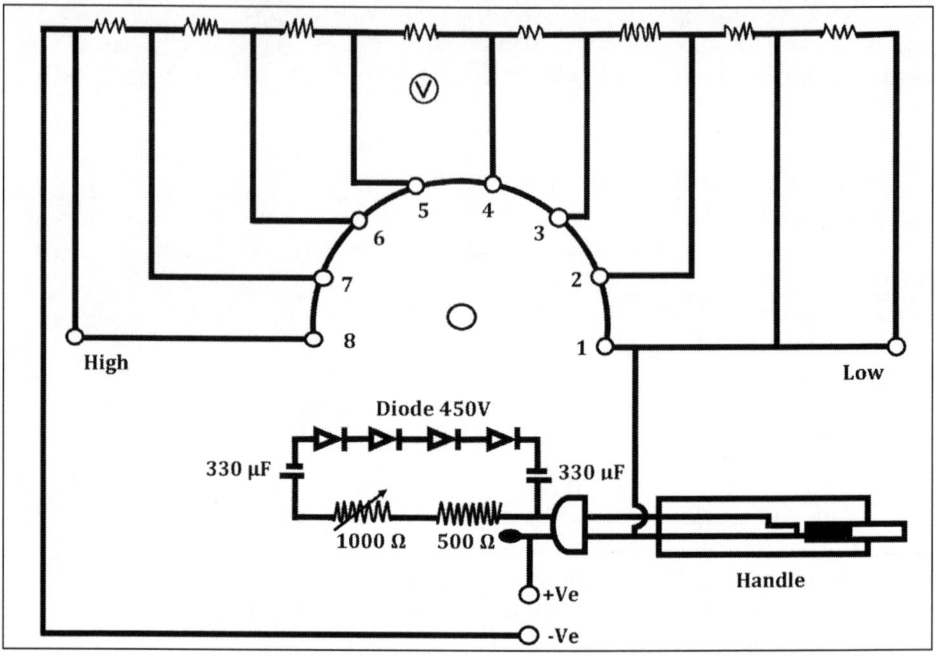

Schematic Diagram of Power Source for Electric Fish Catcher.

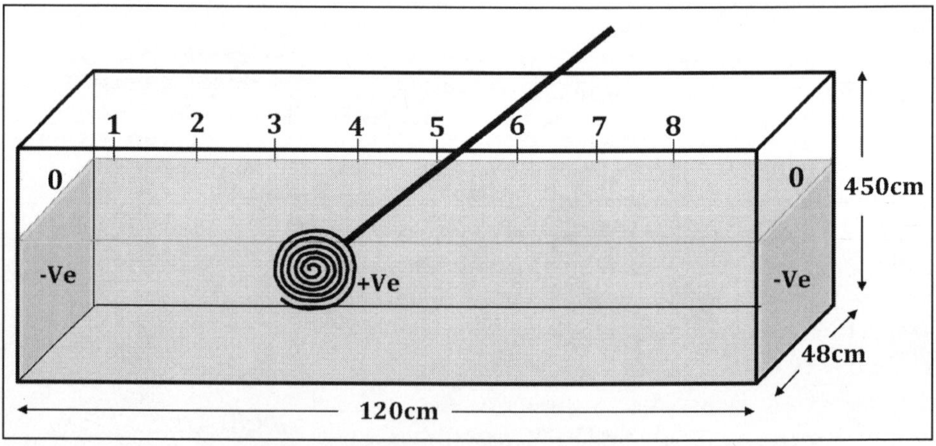

Schematic Diagram of Field Pattern with Emerged Catcher Electrode.

Conditions within the tank during test experiments, produced a homogeneous electric field with an uniform current density. Homogeneity within the electrified field was verified through two probes, one for voltage gradient measurements (V/cm) and the other for current density in μA per mm^2. Specific electrical resistance and ambient water temperature of the experimental tank were recorded by ohmmeter and thermometer, in ohms and Celsius scale respectively.

The distance between the electrodes was maintained at 110 cm except in case of small sized fishes when the distance between the electrodes was reduced to 55 and 27.5 cm respectively.

Two electrical (continuous DC and pulsed DC), two environmental (standing water and flowing water) treatments were applied on eight species of fish individually (*Labeo rohita, Oreochronis mossambicus, Mystus tangra, Trichogaster fasciatus, Notopterus notopterus, Monopterus cuchia, Rhyncobdella aculeate* and crocodile fish). Limited fish availability did not permit application of all the four treatments to all the eight species.

Live fish specimen which were caught in the morning from ponds, were collected in oxygen packing and were kept in four tanks, fed for a minimum period of three days to acclimatize in the test environment. The electrical resistances between different points in a fish body were measured along with the total length and body surface area in respect of four species. After observing the general physiological conditions (gill movement) and normal behavior in the field (coordinated swimming movement), the fish was individually subjected to electrical exposure of rising current intensity and their reactions and the threshold values for initiating those reactions were recorded along with significant behavior during the exposure, both in standing and flowing water (water flow 15 cm/s) both for continuous and pulsed DC. The percentage of identifiable reactions; fright reflex (FR), oscillotaxis (OT), galvanotaxis (GT) and galvanonarcosis (GN), together with the threshold values of current densities required to exhibit those reactions of test fishes were determined.

The data thus obtained from the above tests were synthesized to develop a catcher electrode with suitable electrode arrangements. The efficiency of the catcher electrode was determined in the same tank, except the water depth of the tank was raised to 24 cm with mixed population of all the species combined together. The catcher electrode (anode) is helix coil made of 3 mm diameter copper wire without insulation attached to a wooden handle provided with push button switch. The power supply was from the same inverter used for test experiments and field voltage and intensities were controlled by the rheostat. The voltage gradient and current densities in the tank were determined through the probe electrode at different distances from the immersed catcher electrode. Three variables in the arrangements of cathode were tried. The performance of catcher electrode to influence fish species for FR, OT, GT and GN with respect of three types of cathode arrangements (single cathode of aluminium plate placed vertically; two cathodes of aluminium plates positioned at tank extremities vertically and one cathode of aluminium plate placed horizontally, at the bottom of the entire tank) were observed.

Significant Findings

The reactions of fishes in increasing current density were categorized as:

1. Fright reflex (FR) – Sensing the current with expanding fins, jerks of body, moving backward without changing position and orientation.
2. Oscillotaxis (OT) – Movement between the electrodes to secure a less irritated orientation and a place either perpendicular or 45° degree angle to the field lines of current conduction.

3. Galvanotaxis (GT) – Forced and directional swimming movement to the positive electrode (anode) with head pointing to the anode and stays there without any further movement as long as current flows.
4. Galvanonarcosis (GN) – Either lay on its side or belly up without any further movement till the current flows. Gill movement slowed down in this state and the fish recovers quickly after it is removed from the field and return to its normal swimming position.

Specific reactions of different fish species were recorded in details, so as to consider the degree of their affinity towards the catcher electrode.

L. rohita jumped out of the field of the experimental tank with jerks of head and body tremor when parallel to the lines of current conduction. They also exhibit backward movement from negative electrode towards the anode when parallel to current lines and at times remained perpendicular to the field lines in the mid-field region. Gill movement increased after shock treatment. Fishes invariably jumped out of the field and maneuvered to remain either perpendicular to the field lines horizontally or vertical to the bottom plane. *L. rohita* needed a slightly higher current density for narcosis.

O. mossambicus, when parallel to current lines exhibited jerks of head. Bending of vertebral column, just behind the operculum was observed in smaller fish, unable to keep up swimming position, jumping sideways at the bottom of the tank with fading body color. Immediately after exposure the effected fish died within an hour with bend vertebral column and patches of fading body color after switching off the current. Unbroken fishes always pointed its head to negative electrode and tapped off minimum body voltage by orienting at 45° angle to field lines. Their gill movements were reduced after shock treatment. Most resistant species to electric shocks. It needed not only higher current density and voltage gradient to exhibit narcosis, but also required longer exposure time without break.

M. tangra moved to mid-field with rapid undulation of tail and jerks of head. By bending its body extremities moved to negative electrode and oriented its body axis 45° angle to field lines. All the fishes found to tap off minimum body voltage and exhibited higher gill movements during and after current flows.

N. nototpterus is the most sensitive species to DC. Smallest current density and voltage gradient were required to influence them. Complete narcosis was observed in a much lower voltage gradient and current density than the other species tested. *Notopterus* showed anodic taxis (galvanotaxis) either moving towards anode or in the middle facing the positive electrode, when unable to move out of its own accord as long as current flows. Recovery from current exposure took long time in *N. notopterus* and was directly proportional to the effective period. Its gill movement became slow after the current exposure.

T. fasciatus showed backward movement to positive electrode pointing its head towards the negative electrode. It exhibited circular movement with head towards the bottom and tail up at fright reflex. Bending the body like an arch became motionless, head pointing to the negative electrode. At higher current density, they settled near the positive electrode unable to move, till the current flows. On switching off the current they regained their normal swimming position. Erratic movement in the field during fright reflex and bending of the body, unable to move at higher current densities were the features of all the fishes of the species during flow of current. Regained normal swimming position after the current flow ceased. The species was most sensitive to electric field as all the fishes of this species were narcotized at lower current density created by 15 volts.

R. aculeate exhibited no response while pointing its head to positive electrode except bending of body like "S". But on facing negative electrode, strong irritation of the body was noticed and the fish maneuvered to remain perpendicular to field lines. With the gradual rise in current intensity, except the change of place and bending its body like "S" in higher current density, no other reaction was in sight. In sharp current rise, the fish violently moved between electrodes first, jumped out of the tank, but when put back moved violently to positive electrode, narcosed, head pointing to positive electrode with straight body as long as current flows. Gill movement increased three times of the normal after current exposure. The fish after current exposure invariably bent its body like "S" and remained motionless.

Crocodile fish with expanded fins exhibited jerks. When perpendicular to field lines, avoided positioning parallel to field lines, in continuous current flow. The fish moved between electrodes with raised dorsal fin and finally died at the bottom at 45° angle to the field lines in one case. In pulsed DC fish body vibrated.

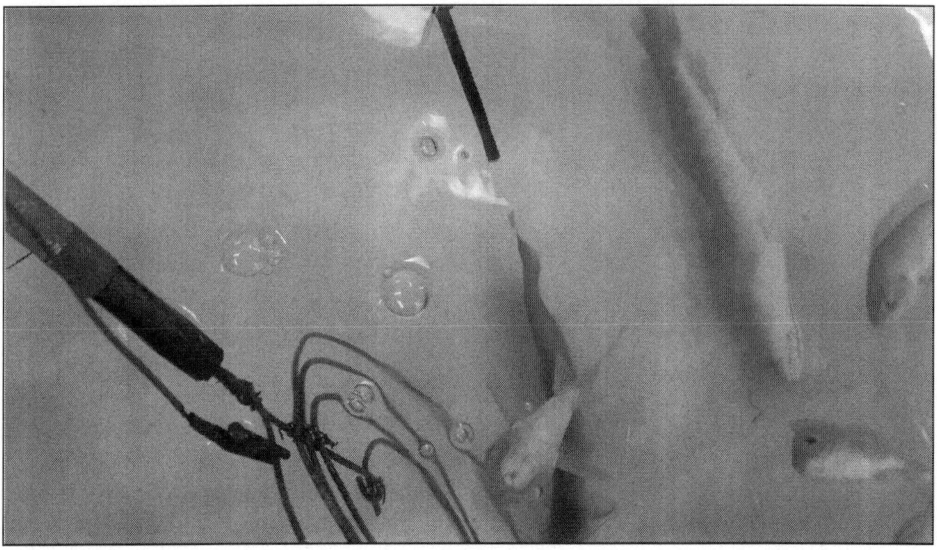

Fishes Attracted to Catcher.

M. cuchia moved violently between electrodes and jumped out of the experimental tank. Their body turned hard and rigid and not pliable as observed prior to electric shocks. Gill movement slowed down. Found to coil and uncoil body with each stimulation in pulsed DC. Fish was found to recover after 10 minutes from narcosis. In heterogeneous electric field, the fish avoided electrified zone and prefer to stay outside the field. But on coming a portion of its body in the electrified area, the fish stood erect on the posterior half of the body, like a snake, with its head towards the water surface. Jerks of the body with each pulse for a few seconds and thereafter violently moved between the electrodes, coiling its body. The fish lied at the bottom in a narcotic state and remained motionless up to 20 minutes. After 20 minutes, the fish again positioned on its posterior end with head up to the water surface for one hour and 45 minutes.

In a large tank with homogeneous field, the fish violently moved between the electrodes, curled its body and after 2 minutes laid flat straight, parallel to current lines, head pointing to positive electrode in a hypnotic state. After stopping current flows, the fish took 5 minutes to regain normal swimming movement.

Claims for the Invention

All the fish (*L. rohita, O. mossambicus* and *T. fasciatus*) jumped violently during fright reflex. At increased current intensities both, *L. rohita* and *T. fasciatus* exhibited narcosis in the middle and near negative electrode being parallel to field lines, while *O. mossambicus* was narcosed near negative electrode at perpendicular position to field lines with bending its body.

When exposed to pulsed DC, *L. rohita* and *O. mossumbicus* remained stationary parallel to field lines, whereas *T. fasciatus* occupied transverse position to the field lines and remained motionless as long as current flows. With increased current intensity of pulses, *L. rohita* and *T. fasciatus* oriented themselves perpendicular to field lines. But *O. mossambicus* either remained perpendicular or 45° angle to current lines with occasional spurting short movements.

Labeo rohita and *Monopterus cuchia* exhibited 50 per cent and 100 per cent galvanotaxis (forced involuntary movement to positive electrode) in current densities of 0.07 - 0.08 and 0.75 - 0.80 $\mu A/mm^2$ respectively followed by cent percent galvanonarcosis in higher current densities in standing water with continuous DC. In pulsed DC, however, *M. cuchia* after fright reflex instead of showing galvanotaxis and galvanonarcosis have undergone oscillotaxis either by standing on its posterior end perpendicular to the bottom plane or by coiling and occupying an orientation of 45° angle to field lines and bottom plane even at much higher current densities, than continuous DC (6.0 – 8.5 $\mu A/mm^2$).

In the flowing water, *L. rohita* exhibited galvanotaxis and galvanonarcosis at a much lesser current densities (0.02 - 0.06 $\mu A/mm^2$) than in standing water, although the percentage of identifiable reactions were more discrete in standing waters. The occurrence of galvanotaxis and galvanonarcosis in *L. rohita* was less pronounced in pulsed DC in the flowing water and higher current densities were required to initiate these responses (0.25 - 0.28 $\mu A/mm^2$ in pulsed DC 56/min and 0.075 - 0.09 $\mu A/mm^2$ in pulsed DC 120/min).

The performance of the catcher electrode (anode) which was developed to produce attracting effect (galvanotaxis) was tested with different electrode arrangements especially of the cathode to increase the effective zone.

The arrangement of the positioning positive and negative electrodes in tank extremities, as done, in case of tests in homogeneous field did not influence majority of the fish population (mixed population) for higher percentage of galvanotaxis followed by narcosis. Fishes after fright reflex have undergone oscillotaxis (8 – 20 per cent) in some cases.

Placing the negative electrode (aluminium plate) covering the entire bottom of the tank also did not show better response of fish for galvanotaxis followed by narcosis and 50 percent of fishes after fright reflex have undergone oscillotaxis.

In three sets of experiments where two negative electrodes (aluminium plate 45 X 40cm each) positioned vertically at both the extremities of the tank, 90 – 100 percent fishes exhibited galvanotaxis and narcosis near the catcher electrode (anode) at current densities of 0.00025 – 0.001 $\mu A/mm^2$.

With the object to develop an effective electrical fish catching unit, tests were carried out on eight varieties of Indian fresh water fishes to determine threshold current densities for showing different responses in homogeneous electric field. In nineteenth and twentieth centuries, several authors of Europe, America and USSR investigated the electrical effect on different fish species in different environments. The results obtained by them were also different. Not much work has been done with Indian fishes except Biswas (1971a, 1976, 2015).

The fright reflex (FR), oscillotaxis (OT) to tap off a minimum voltage between body extremities, galvanotaxis (GT), unidirectional forced movement toward positive electrode and galvanonarcosis (GN), a hypnotic condition with slackness of body near the anode are some of the major categories of fish responses observed in electrical field, by most of the investigators.

In the present invention, stress was given to identify the above mentioned reactions with respect to Indian fresh water fish species and the field intensity required to initiate those reactions in a homogeneous under water electric field. The significant reactions exhibited by the members of each species were taken into consideration along with the threshold values of electric parameters in order to design the electric fish catcher.

Besides the specific variations in threshold values of current densities required for different reactions, lengths of the fishes have been found to influence the threshold values (Biswas. 1971b, 1976) as larger the body extremities (lengths) tapped off higher potential differences and the fishes had the chance to be influenced more easily in a smaller current intensity. In the initial experiments specific variations on the threshold values initiating the reactions and the percentage of fish exhibiting those reactions were taken into account both in standing and flowing waters in continuous and pulsed DC of low frequency.

The non-availability of commercial DC power source and impulse generator (Dolan and Miranda, 2003) an inverter was used which converts AC to step down DC voltages in eight steps. An AVO meter was used to measure electrical parameters during the tests.

No significant differences have been observed in threshold values of the reactions in standing and flowing water except higher percentage of galvanotropic responses in standing water. The flow of water (15cm per second) might have over ridden the galvanotropic response of fish.

The low pulse rate (56 – 120 per minute, due to low capacity relay unit) was found to be ineffective to bring out galvanotropic responses in fish even in higher current densities probably due to minimum effective periods of current flow in low pulse rate (56 – 120/min).

Electrical method of collecting fish in shallow waters have been developed quite a long back (Benjamin and Gillespie,1966), because they are more convenient and efficient compared to nets and traps. Qualitative evaluation on and comparisons of electro-fishing results by different authors are almost impossible, because varied shocks and shock producing equipments have been used in waters of varying electrical resistivity on different species and sizes of fish.

In the present experiments, multi–species fish population of different sizes were used to study the proficiency of catcher electrode (a helix coil with a handle) to initiate the galvanotropic reactions of fish with different electrode arrangements.

The position of positive and negative electrodes vertically at the tank extremities failed to influence the fishes of mixed population to exhibit higher galvanotropic response to catcher electrode, probably due to the larger distance between the anode and cathode. As a result, many fishes exhibited oscillotactic response.

Benjamin and Gillespie(1966) suggested that the electrodes should be operated close together for better catching efficiency specially in resistive waters.

Lenon and Parker (1955) reported that the pretty type alternate polarity electrode system of original and modified design, has been successfully tested in Park streams having resistivity of 32000 to 117000 ohms/cm^3, whereas electrode system of conventional designs proved ineffective. They however, used the alternating current power and a battery powered back pack unit which provided 240 and 250 V respectively.

Placing the negative eletrode at the bottom of the tank horizontally throughout the entire length, was also not found to be that much effective to bring out galvanotropic response of fish and narcosis. The negative electrode placed horizontally at the tank bottom leaves a little space between the negative and the catcher electrode vertically at the place where the helix coil of the catcher electrode was put into the water. Only that area (approximately 12-14 cm diameter) produced higher voltage gradient and current density (55 V and 0.0025 $\mu A/mm^2$, whereas fishes in rest of the area (60 cm away from the helix coil) were not properly influenced by threshold voltage gradient and current densities (35 V and 0.0005 $\mu A/mm^2$) and fishes did not show galvanotropic responses, instead they were found to occupy perpendicular position to the current lines and bottom plane after getting initial shock to tap off a minimum voltage (oscillotaxis) between their longest body extremities.

Position of the negative electrode with respect to anode was found to play a major role in creating threshold voltage gradient and current density for galvanotropic reactions of fish, which is most desirable in electrical fishing.

So when two negative electrodes were positioned at the longest extremities of the tank, the field pattern changed to an optimum level (0.001 – 0.0025 $\mu A/mm^2$) and cent percent fishes exhibited forced swimming to the catcher electrode (helix coil) and undergone hypnosis around the catcher electrode.

Justification for the claim

Comparative study of some of the prior arts related to an electric fish catcher

Pub. No. US 3693276 A

Electric sea fishing apparatus relates to a device by means of a suction pump. The fish being attracted by a lighting lamp and directed towards the intake of a suction pipe by the effect of a pulsating electric field serving as a relay zone between the attracting light radiation of the lamp and the hydrodynamic suction field of the pump. The invention relates more particularly to the means of producing a pulsating electric current in the volume of water close to the intake of the suction pipe and to means for suitably distributing the electric field in this volume.

The dimension and the relative position of the electrodes (one anode and two cathodes) have been determined in such a manner as to obtain a greater volume in which the electric field varies less rapidly while remaining above the galvanotactic threshold, and the intersections of this volume with the light attraction and hydro–dynamic capture zones respectively must be large enough to ensure the easy transfer of the fish. Effective on deep–sea fishes such as sardines, sprats, anchovies, mackerel and herring.

The device, the electric fish catcher, is not meant for sea fishing in conjunction with lighting lamp and a suction pump as patented in publication number US3693276A.

In the electric fish catcher for which patent was applied produce a wide effective zone of electric field in the water, where fish of all varieties were attracted to the catcher electrode, immobilized and caught. The catcher is meant to be operated in fresh water rivers, streams, reservoirs and will be effective on all types of Indian fresh water fish species.

Pub. No. US 005797210

Comprising a trawling net and a combination of electrodes, which during operation are towed over the bottom. The electrodes are towed over the bottom, ahead of the net at a distance in front. A pulse generator which is connected to the electrodes to generate during operation a pulsed electric field in the water ahead of the trawl net.

The electrodes (three mentioned in the write up) one connected separately to the pulse generator, whereby the water contracting surface of each electrode is limited such that specific electrical resistance is created. The pulse generator is embodied such that each of electrodes can be powered separately to generate a field between momentarily powered one electrode and another electrode.

Effective on bottom dwelling and off bottom sea fish.

The device, the electric fish catcher is entirely different from the patented publication no. US005797210, where three electrodes are towed, ahead of an opening of a trawl net which was towed at the bottom of the sea. The electric fish catcher is an effective means of catching fish in fresh water streams, rivers and reservoirs by creating an underwater electric filed, where fishes are attracted to the catcher electrode and immobilized, since the galvanotactic threshold is maintained around the catcher electrode around 4 feet when operated at 110 volts and around 8 feet when operated at 210 volts.

Pub. No. CN2116349(U) – Electric attracting fishing machine

Comprise an alternator composed of ordinary three phase asynchronous motor and an excitation capacitor connected in parallel with its output line. The generator is driven by an engine and the output line is connected to a cathode placed in pool water, and its positive terminal is connected to the anode operated by fisherman. The number of anodes is preferably two or more to ensure that at least one anode remains in the water during fishing. The voltage between the anode and the cathode is maintained at 400 - 500 volts.

Effective in large and medium sized reservoirs and professional fish ponds as especially for catching ground fishes in dense, bumpy river and lakes.

The device, the electric fish catcher is something different from electric attracting fishing machine, patented under CN2116349 (U). In the electric attracting fishing machine, they have used a very high voltage of 400 - 500 V which is too much accident prone and can be dangerous to the operator or to persons remaining nearby. They have used one cathode and two or more anode (2 or more), as a result they had to increase the high input voltage (400 – 500 V) to create the threshold intensity which stands for a high risk factor.

In the electric fish catcher (for which patent application was made) one anode and two cathodes were used to widen the effective zone of fish capture in a low input voltage (110 V for distance of 4 feet and 210 V for a distance of 8 feet from anode) which are less accident prone than electric attracting fishing machine.

Pub. No. US4970808(A) – Electro – acoustical fishing lure

The introduction of the electrical pulses and sonic waves into the water habitat of the fish is accomplished by suspending electrodes of different electrode potential attached to the fishing lure. The electrodes are configurated to flutter when drawn through the water and the fluttering causing the electrodes to make momentary contact. The electrodes of different potential and the water as electrolyte constitute a voltaic cell which discharges an electrical current pulse when the electrodes make contact. The impact of the colliding electrodes simultaneously generates an acoustical wave in to water; it has been long known that electrical pulses and sonic waves attract fish/No indication has been made about the intensity and frequency of pulses and the place of operation (sea or fresh water) and the type of fish which were attracted.

The electric fish catcher for which patent application made is entirely different by the way that the catcher directly catches the fish by creating and maintaining galvanotactic threshold in the water around fish population without the aid of net or any other fishing implement.

The electro-acoustical fishing here patented under US4970808(A) on the other hand is a device where electrodes flutter, causing to make momentary contact creating a condition of a voltaic cell with the water as electrolyte and an electric current pulse is discharged. The colliding electrodes, claimed to generate acoustical waves. This is a passive method even if the fish were attracted and maintaining galvanotactic threshold.

Pub. No. JP2016-015900 – Electrical float

Provided an electrical float that enables a fisherman to prepare for main bite by informing of state of pre-bite upon biting so as to know position of float by informing of the float landing on water when the fisherman casts the float into a river or sea water and further enables catches to be increased by informing of the main bite.

Meant for angling – fishing at night.

Cannot be compared with the present invention of an electric fish catcher since fishes are directly attracted to the catcher and caught.

The electrical float (JP2016-015900) is an aid to do angling at night, where fisherman by observing the changes of lighting conditions of a light emitting diode can detect a pre-bite state, enabling the fisherman to prepare for the main bite.

An indirect way of improving the fish catching efficiency by angling a method of sport fishing.

Comparing with previous patents in the area and summing up the differences identified in the present invention are,

Direct current was used, which caused the fish to swim involuntarily to the catcher electrode (anode), where all the different fresh water Indian fish species were immobilized and caught.

Multiple electrodes array of 2 cathodes and 1 anode (unlike one cathode and two or more anodes), of electric attracting fishing machine under patent no. CN2116349(U), increased the width of the effective zone and maintained galvanotactic threshold in a lower safe voltage of 110 – 210 V for fish capture.

The helix coil catcher attracted fish from greater distance, deeper water and hide outs easily.

Problem Solved by the Invention

Any electro-fishing system requires some minimum effective value of current density produced by the electrodes. The minimum value will vary with water conductivity, temperature and the target fish to be caught. Direct current (DC)

field on reaching the effective value, will cause the fish to swim involuntarily to the positive electrode, where they are immobilized when they reach the strong field near the electrode.

But in the present electro-fishing practice, generally two electrodes (one positive and one negative) is used. The electric field thus produced between the two electrodes is sometimes unable to make wider effective zone and strong enough to influence the fish for anodic attraction in low voltage applied to the electrodes in order to avoid unnecessary fish mortality and electrical accidents.

Multiple electrode arrays (one anode and two cathodes) has been utilized in the present invention to increase the area of the effective zone of a DC field with low input voltage, thus saving power requirement and a better safety measure to the operator.

The invention has also quantified the field pattern of multiple electrode array together with the reactions of mixed fish population (different fish species encountered in the inland waters of India) with emphasis on desirable attraction effect of fish to positive electrode (the catcher electrode) for bulk fish capture without net in low voltage (110 V) DC.

The proto type power pack and the multiple electrodes developed by the inventor is capable to supply 3.5 kW, 110 volt rectified DC from commercial power supply of 220 volt AC or from portable AC generator.

Solutions to Solve the Existing Problems

Solutions to the existing problem was to either increase the size of electrodes, especially, the cathode and the use of higher voltage, or to use pulsed DC. Too much increase of cathode size involves the unwieldy of the gear and will cause physical obstructions during operation. Higher input voltages to the electrodes stand the risk for electrical danger to the operator, particularly, to back pack waders.

As regards the use of pulsed DC, each fish species has been found to respond to a specific pulse frequency for attracting effect. When the frequency of the pulse is matched to the natural wriggling frequency of the fish, it will cause the fish to swim up to the positive electrode. In a mixed fish population intended for capture, such an universal/equipments to deliver multiple pulse frequency at a time is not possible to develop.

If the electrical combination, that is, level of voltage, pulse width and pulse rate, is not adequate, fish will immediately escape to the side or towards the negative electrodes.

Solution Provided by the Invention

Direct current was used which will cause the fish to swim forcefully to the positive electrode, where they (all fish species) were immobilized on reaching the strong field near the electrode.

Multiple electrode array has been developed, which increased the width of the effective zone (area of operation), where fishes of all varieties, possessing neuro-muscular system were attracted to the catcher electrode. Instead of present system of 1: 1 positive and negative electrode, 1: 2 positive and negative electrodes were used. Two 18 – gauge aluminium plate of 29 X 29 cm served as negative electrode whereas the catcher electrode (anode) consisted a 19 cm dia helix coil of 3 mm diameter bare copper wire attached to a wooden handle provided with a push button switch.

The power source consisted of a rectified DC unit of 110 volts obtained through a step down transformer out of 220 volts commercial AC line. Two 330 micro-farad capacitors were used to smoothen the DC output. In the external circuit two variable resistances of 1000 and 500 ohms were used to adjust the output voltage.

Advantages

1. Rectified DC from commercial AC or portable AC generator can be easily obtained at a low cost than pulsed DC with multiple pulse frequencies and pulse width.
2. Multiple electrode array system and their rational placement increased the effective zone of fish capture.
3. The helix coil catcher attracted fish from greater distance, deeper water and hide outs easily.
4. Lower voltage (110 V) could influence the fish for directional forced swimming due to adopting the multiple electrode array system, saved power and risk of killing fish and electrical accidents to the operator.

3

M and K Technique of Harmless Electric Fishing

Conceived and Developed with Trials – Dr. K. P. Biswas

Associates in the Work – Miss Namrata Basu, Miss Arpita Roy, Mr. Amitavo Roy and Mrs. Nabanita Chakraborty

Background of Invention

Harmful effect and injury of fish due to electric fishing has been a matter of concern for some time past among the investigators working on electric fishing techniques. Experiments on the physical effects of electricity on fish have been carried for some time. Hauck (1949), first to investigate the internal effects of electric shock on rainbow trout (*Salmo gairdnerii*), found incidences of fractured vertebrae, curved spines, ruptured arteries and veins, haemorrhaging and tissue death. Subsequent studies have confirmed and attempted to quantify these effects. The most common internal injury caused by electro-fishing is haemorrhaging. Schreer et.al (2004) found that in groups of rainbow trout subjected to different voltages, frequencies and shock duration, haemorrhaging occurred in 29-100 per cent of fish showing that even the lowest voltages and frequencies caused a substantial occurrence in internal haemorrhaging. Spinal injuries can also be common during electro-fishing activities. While it was initially thought that the severe muscular contractions experienced during tetany were responsible for the compression fractures, often seen in fish exposed to electric field. Sharber *et al.* (1994) have shown that rather than the higher voltages which cause tetany, spinal injuries are associated with higher frequencies and can occur at any point after the onset of galvanotaxis. Besides from these recorded physical effects, electro-fishing is reported to cause physiological changes in fish, when exposed to an electric current. Oxygen consumption of rainbow trout increases rapidly to 150 per cent of resting levels with AC, 130 per cent with DC and 110 per cent with pulsed DC and can take from 30-120 minutes to recover to pre-shock levels (Emery, 1984). This is possibly as a result of lactic acid built-up due to rapid muscular contractions induced by the electricity. While most fish will recover from this build-up of lactic acid within 4-12 hours, some fish will never recover resulting in delayed mortality (Emery, 1984). Electric shock also has some effect on cardiac functions. Schreer *et al.* (2004) noted that rainbow trout suffered cardiac arrest when they encountered an electric field which lasted for long duration of shock. Immediately after shock, there was a period of arrhythmia which lasted for a few seconds to several minutes. While heart rate increased to only 108-132 per cent of resting values and took 40-114 minutes to return to normal. With regards to cardiac functions shock duration appears to be the major factor, while higher voltages and frequencies result in longer recovery time (Schreer et.al., 2004). It is important to note that level of response varies between species and individuals. The studies reviewed were all carried out on Salmonids, which are species, recognized to be particularly sensitive to electricity (Snyder, 2003). However, similar responses have been seen in other species of fish. (Adams *et al.*, 1972; Dolan and Miranda, 2004; Henry *et al.*, 2004)

Biswas *et al.* (2019) observed the occurrence of spinal injuries, like curved spine in *Labeo rohita* and *Cirrhinus mrigala* (3-7 per cent) when subjected to DC and low frequency pulsating DC of 65 to 80 volts. Delayed mortality has been observed in 43 per cent of injured fish even after six days of electric exposure

Indian freshwater fishes, in an electric field, was frequently observed to frantically searching for a hideout in non-effective zone, where they can stay comfortably. Even if they were physically removed and placed in the effective zone, they again took shelter in non-effective zone. This behavior of fish have been observed in most of the Indian freshwater fish species. It was, therefore, thought that if a non-effective tunnel or hole could be created in the electric field, near positive electrode, where fishes could be induced for forced involuntary swimming towards the anode. On reaching nearer and nearer to the anode, they will experience higher current intensities restricting their escape from anodic attraction. If in this situation, fishes could find a tunnel of lesser current intensities leading to a non-effective zone, fishes will hurriedly rush towards the low intensity gap and then to the non-effective captive zone.

After conceiving this idea, the task remained to design and develops an effective electric field with a low intensity tunnel and a non-effective captive zone.

The Invention

The design under this situation centered round the use of set net, fortified with positive electrodes in front of the net and placing a negative electrode, as far as possible from the positive electrode, so as to create an effective electric field. The effective electric field, in front of the positive electrode, will be provided a gap of lesser current intensities (a blank hole) by making suitable arrangements of anodic system.

The history of the development of the set net technology (a type of trap net) centered around two conflicting objectives; (a) to make a trap net, which is easy for the fish to enter and (b) to frame a net from which it is difficult for the fish to escape once they have entered. In conventional type, a funnel shaped entrance is provided for easy entry of fish swept off by the water current. While the small funnel opening to the bag net supplement the function of bag net by luring the fish in and then cut off their rout to escape. The bag net gather fish and prevent them from escaping.

At the set net fishery, the gear is fixed in some specific areas for a long time, so as to get fish or crustaceans regularly every day, when they come into the bag net unconsciously and is known as " wait and see " fishery. Several natural factors, such as, bottom topography, tidal and stream conditions influence the set net fishery. It takes advantage of the movement of fish and shell fish with the tides in shallow sea areas, which are changed into an expansive tide land at high tide. Most set nets need to be set in areas with relatively mild water current of 0.4 knot to 0.6 knot per hour. As such set net is not successful in lentic or standing waters, like, ponds and tanks, reservoirs, lakes *etc.*

So, to make the set net functional in standing waters, the idea of anodic attraction of fish was thought of by creating an effective field in a section of the water area with suitable electrodes arrangement, first to attract fish of effective zone towards the positive electrode and then to make tunnel of perception zone, which will lure fish to escape from the effective field through a funnel net into the bag net, where no electric field will exist. Fish will thus, influenced by electric field will pass through the perception zone in front of the anode, unharmed into the bag net and caught. The strong electric field surrounding the perception zone near anode will attract fish electrically towards the anode.

An electrical set net was designed, developed and tested for its efficiency in catch per unit effort (CPUE) with respect to non-electrified conventional set net.

A framed set net with wide opening and narrow rear having two wings attached vertically in front of the net, to give a wide coverage was designed as per following specifications.

Net Proper

4 mm GI rod was suitably welded to give a trapezium shaped frame of the net. The front opening of the net was 27 X 23 cm, while the rear opening 23 X 21 cm. The entire net excepting the front and rear was covered with synthetic net of multifilament twine with a mesh size of 1 cm in stretched condition. The rear end of the net was provided with a pouch of same webbing to collect fish which will enter the net.

Wings

Two wings of 38 X 23 cm mounted on 4 mm GI frame of same netting materials (synthetic twine 210/6/4) and same mesh size were attached to the two sides of net mouth vertically and hinged with the vertical frame of the main net, so that the wings can be stretched sidewise to cover a more area to guide fish into the net.

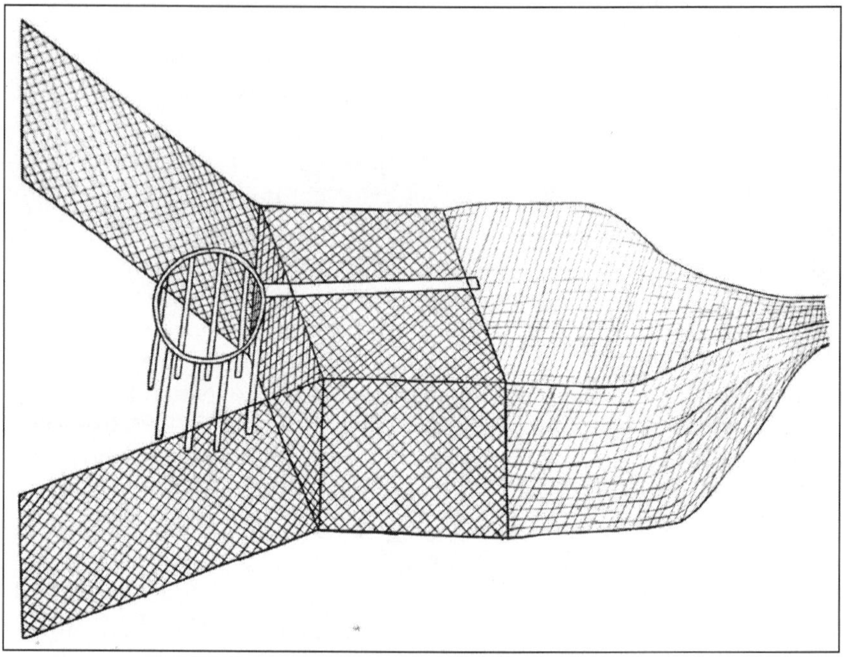

Schematic Diagram of Electrified Trap Net.

Electrodes

The anode or positive electrode consist of an aluminium ring of 14.5 cm dia. around which 8 solid aluminum rods of 20 cm length and 5 mm thickness hang loosely on copper wire stirrups making them electrically connected with the aluminum ring, which was connected to the positive terminal of the power source.

18 gauge aluminum plates of 30 X 28 cm and 101 X 56 cm were used as.negative electrodes in small and large experimental tanks, which were vertically placed in the furthest extremities of the experimental tanks of 121 X 30 cm and 174 X 103 cm at a distance of 79 cm and 132 cm from the positive electrode.

Experimental Tanks

Trials with the electrified set net were conducted in two glass tanks of small (where length of the tank was proportionately larger than width, 121 X 30 cm); and the other large tank (where length of the tank was almost double the width, 174 X 103 cm). The smaller tank was filled with water to a depth of 22 cm and the bigger one to a depth of 44 cm at the time of trials.

Power Source

The power source consisted of one 1.5 kVA inverter which stepped down and rectified 220 volt AC to DC with an output voltages of 40 to 200 volts at eight steps.

The output DC voltages were further reduced by three variable resistance (rheostat of 1000 and 500 ohms) and could be suitably increased whenever necessary to increase the field intensity. A volt meter in the circuit indicated the voltage applied to the electrodes and the density of current in the water of the experimental tanks were determined with the help of a probe and a milliampere meter. Direct current and oscillating DC of low frequencies were used during trials. An oscilloscope was used to check up the pulse shape and oscillating frequencies (Hz).

Tests were carried out in two glass tanks, one small and the other big tank with two different experimental setups, where the electrodes of different sizes and shapes were taken for the trials. 18 gauge aluminum plate of 30 X 28 cm was used as cathode for the small tank along with the aluminum ring and 8 droppers, made of solid aluminum rod hung, loosely around the droppers were used as anode to create an electric field in the small tank to know the reactions of different varieties of freshwater fishes during the trial. Placing the cathode vertically at the longest extremities of the small tank and suspending the aluminum ring with droppers in front of the set net in such a way that the trap net along with the aluminum ring with droppers remain immersed under the water of the small tank. Aluminum droppers were suspended around the ring leaving a gap of 11 cm in front of the trap net opening, 4 droppers at each side separated by 3 cm from each other, so as to create 11 cm wide perception zone (escape tunnel) at the center in front of the set net. Four droppers at each side established a strong effective zone to force fishes to swim towards anode when the current flow to the electrodes was switched on.

The second trap net was 47 X 36 cm with two wings of 47 X 36 cm in each side and the opening of the trap net being 47 X 36 cm. Negative electrode consisted of 18 gauge aluminum plate of 101 X 56 cm, which was kept in the longest extremities of the big glass tank. Seven aluminum tubes of 21 cm long, hanging loosely over 42 cm aluminum tube of 12 mm diameter, of which one was kept at the center of the trap net opening and the rest six, three each at one end with spacing of 0.5 cm apart from each other. Three electrodes at each side were 16 cm apart from the central one. A steel ball of 6.57 cm diameter was hanging inside the trap net at the rear end of 26 cm from the hanging droppers at the front line.

Instrumental Setup for Harmless Electric Fishing.

The power source used in the trials was from the alternating current (AC) service line through an inverter and an oscillator, which oscillated the DC input voltage between one to six per second. Before applying to the electrodes of the experimental tank, the oscillating voltage from the oscillator passed through variable resistance of one 1000 ohm and two 500 ohm each. The input voltages were measured with the help of a voltmeter connected in parallel to the supply line of the electrodes. The field strength (current density) was measured with the help of a probe and a milli-ampere meter and was expressed as mA/sq. cm.

In the small glass tank, aluminum plate (30 X 28 cm) as a cathode and aluminum ring with droppers as anode, fitted in front of the small trap net, were used at a water depth of 22 cm. The droppers were so arranged (4 droppers at each side) so as to leave a gap of 11 cm at the center in front of the net mouth, which will act as escape channel for the fishes into the net.

Seven varieties of Indian freshwater fishes and one variety of freshwater prawn and an estuarine crab (mud crab) were taken for the behavior studies in the electrical field and if they can be captured in the trap net at different population densities both for single species and multi-species combination in small tank with small trap net and in big tank with small and big trap nets with different types of anodes in order to find out the best one for attracting and trapping fishes in the net behind the anode unharmed.

Species like, *Labeo rohita, Oreochromis nilotica, Labeo bata, Clarias batrachus, Cyprinus carpio, Channa punctatus, Mustus tangra, Mystus cavasius, Labeo calbasu, Channa striatus*, prawn (*Macrobrachium rosenbergii*) and a crab (*Scylla serrata*) were introduced in the tank after recording their length at different population densities for single species and multi-species combination.

Direct current (DC) and oscillating DC of different frequencies, one type at each time was raised slowly to increase the electrical strength in the tank water in a such way, so as to exhibit the involuntary movement of the influenced fishes towards the positive electrodes, passed aside the droppers and took shelter in the trap net behind the anode. The applied voltage between the electrodes and current densities established between the electrodes, when the fishes exhibited strong swimming movement towards the anode was recorded through the voltmeter, connected parallely between the electrodes and by the milliamperemeter through the probe. The exposure to the electric field was stopped immediately after fish entering into the trap net and the time of exposure was recorded. The movement of fish and taking refuge in the trap net was recorded video graphically for each trial after calculating percentage of their moving towards the positive electrode and entering into the trap net behind the anode. Trials were carried out both in small and big tank in different water conductivities applying different types of current in different population densities.

In DC reaction of fishes in gradual rise of current intensities and sharp rise of current densities were compared for their percentage of capture in the trap net.

Trials were first conducted in the small tank, when the small net was first kept under water at one end of the tank. The water level of the tank was so maintained

(22 cm) so that the net remained completely under water. Individual fish, two or four in numbers of each species, at a time, or a combination of multi-species were introduced in the tank containing trap net for 15 minutes, in each case, so as to observe if any fish enter into the trap net. When individual species of carps, catfish, murrel, tilapia were introduced into the tank, except one or two murrels, fishes of other species were found to avoid the trap net. The murrels (3-6 per cent), who sought a hideout entered into the trap net when multi-species fish population were introduced in the tank. No other fish species entered the trap net.

On placing the anode (aluminum ring with 8 droppers) in front of the trap net and a 18 gauge aluminum plate cathode at the opposite end of the tank vertically, completely immersed under the water, and connecting the anode and cathode with the positive and negative terminal of power source, an electric field was created inside the tank water, the strength of which depended on the input voltages at the electrodes.

Before starting the trials with electric field, the temperature and electrical conductivity of the tank water was measured with thermometer and digital conductivity meter.

Individual fish, after measuring their total length and weight to the nearest mm and gram, was released in the experimental tank and left for 10 minutes to acclimatize with the new environment. Its gill movement was counted by counting the opercular beats per minute in order to compare with after shock gill movement to find out the amount of stress the animal suffered on account of current exposure.

Field intensity was slowly increased by raising the potential difference between the electrodes, which was recorded by a voltmeter, connected in the circuit. The density of current flowing through the water during the exposure of current was also measured with a probe and was recorded by a precision milliampere meter. With the rising current density to a threshold value, fish started feeling the presence of electric field in the water, which was observed by expansion of fins, tremor of body and at times jerk of head. The input voltage and current density at the onset of first reaction was noted and the input voltage was carefully increased till the fish was found to swim involuntarily, speedly towards the ring with droppers (positive electrode) and entered into the trap net, pushing aside the swinging droppers, since no electric field persisted inside the trap net.

Slowly increasing DC and DC sharp rise at pre-set voltages were used, besides oscillating DC of 1 to 6 Hz and the reaction of fish at each current type was recorded video graphically, especially with respect to their strong involuntary movement towards the anode and taking refuge in the trap net behind the anode.

No harmful effects or injury to fish was observed.

In the similar manner, experiments were conducted in the big glass tank with the small net and the big net with two different anodic set-ups (one ring with 8 droppers and the other on aluminum conduit tube with 7 droppers of aluminum conduit)

After ascertaining the heterogeneous electric fields that have been created by two different anodic set-ups, attempts have been made to open a window between the fields, where there will be practically little or no influence zone of the electric field. Care was taken to obtain this no influence zone in the vicinity of the positive electrode, so that the attractive effect of the anode can assemble fish entering into the electric field near the positive electrode. Fish on reaching involuntarily near the anode, in a frenzy of getting relief from the tension of the electric field will find the window and will promptly cross the window, enter into the trap net placed behind the window.

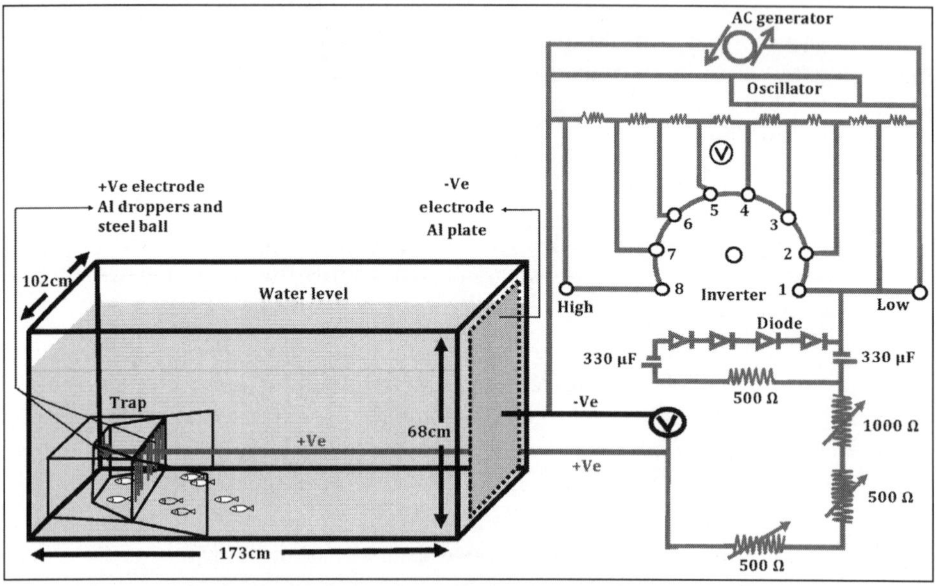

A Schematic Diagram Showing the Electric Circuit and the Position of Net in the Water during Operation.

Significant Findings

The exposure of electric shock lasted between 12 to 164 seconds.

In case of *L. rohita* of 130-199 mm length, in DC, 25-140 seconds: in ODC 1 Hz, 34-97 seconds; in ODC 3 Hz and ODC 6 Hz, 10-97 seconds of exposure were given to fishes. The corresponding input voltages were, 32-86 V, 50-114 V, 22-32 V and 32-60 V respectively. Hundred percent fish moved towards the positive electrode and entered into the trap net.

In a wider area (big tank), similar to the small tank, though 100 per cent *L. rohita* of 140-199 mm length responded for galvanotaxis and entered in the trap net behind the positive electrode, yet there were differences in applied voltages and the time of exposure to complete the capture (42-100 V in DC, 80-170 V in ODC 1 Hz, 45-62 V in ODC 3 Hz and the corresponding exposure time was 42-130 seconds, 90-152 seconds and 112-130 seconds respectively).

L. bata of 150-189 mm length, when subjected to an input voltage of 34-70 V in DC; 55-125 V in ODC 1 Hz, 22-60 V in ODC 3 Hz and 30-40 V in ODC 6 Hz in small experimental tank for a period of 32-160 seconds in DC, 10-117 seconds in ODC 1 Hz, 55-78 seconds in ODC 3 Hz and 35-50 seconds in ODC 6 Hz, all of them were attracted to the anode and took refuge in the trap net behind the positive electrode.

In large area (big tank), though 100 per cent of them exhibited forced movement towards the anode and entered in the trap net behind the positive electrode, where no electric field existed, yet they needed higher input voltages of 62-90 V in case of DC and 55-140 V in ODC 1 Hz, though there was not significant difference in exposure time to initiate and complete those reactions. *O. niloticus* of 150-189 mm size range in DC field partially (33-66 per cent) responded for galvanotaxis and entered in to the trap net at 120-130 V within 9-13 seconds. Though 100 per cent of them responded in ODC 1 Hz for swimming to anode and to the net, but that was possible when the voltage was increased to 160-200 V with an increased period of treatment by 51-90 seconds. Increasing the pulse frequency to 3 Hz and 6 Hz, the threshold values of input voltages were 20-85 V and 72 – 140 V, the exposure time being 26-117 and 48-95 seconds respectively.

Stimulated Fish Entering the Trap Net.

The scale less catfish, *C. batrachus* of 200-209 mm exhibited 100 per cent taxis to the positive electrode and took shelter in the trap net at 120-130 volt within a period of 9-13 seconds. Stimulating them with ODC 1 Hz, 3 Hz and 6 Hz, all of them swam rapidly to the positive electrode and into the trap net in much lower voltages of 30-50 V, 25-42 V and 20-60 V respectively within 19-130 seconds indicating their susceptibility towards oscillating direct current than of DC.

In DC field, both in narrow and wide area (small and big tank) *M. rosenbergii* of 130-199 mm total length (tip of the rostrum to the end of telson) did not respond for galvanotaxis and entry into the trap net even a potential difference of 40-100 V were applied between the electrodes for 102 seconds. Fifty percent of them moved towards the anode and into the net in small experimental tank at 70-120 V, when exposed for 120-180 seconds. In expanded area (big experimental tank), none of them moved to anode and into the trap net, though they were subjected to higher voltages (110-200 V) and for a longer duration (135-249 seconds).

In ODC 3 Hz, hundred percent *M. rosenbergii* exhibited forced swimming to the positive electrode and entry into the trap net at a much lower voltages of 18-30 V and 35-40 V within 8-78 and 122-174 seconds respectively both in small and large experimental tanks.

Basing on the inventory results of individuals of each of the following five species, namely, *Labeo rohita, Labeo bata, Oreochromis niloticus, Clarias batrachus and Macrobrachum rosenbergii*, multi-species reactions of following nine species, namely, *Labeo rohita, Labeo bata, Cirrhinus mrigala, Oreochromis niloticus, Notopterus notopterus, Cyprinus carpio, Clarias batrachus, Heteropneustis fossilis* and *Channa punctatus* together was studied in wider area (large experimental tank).

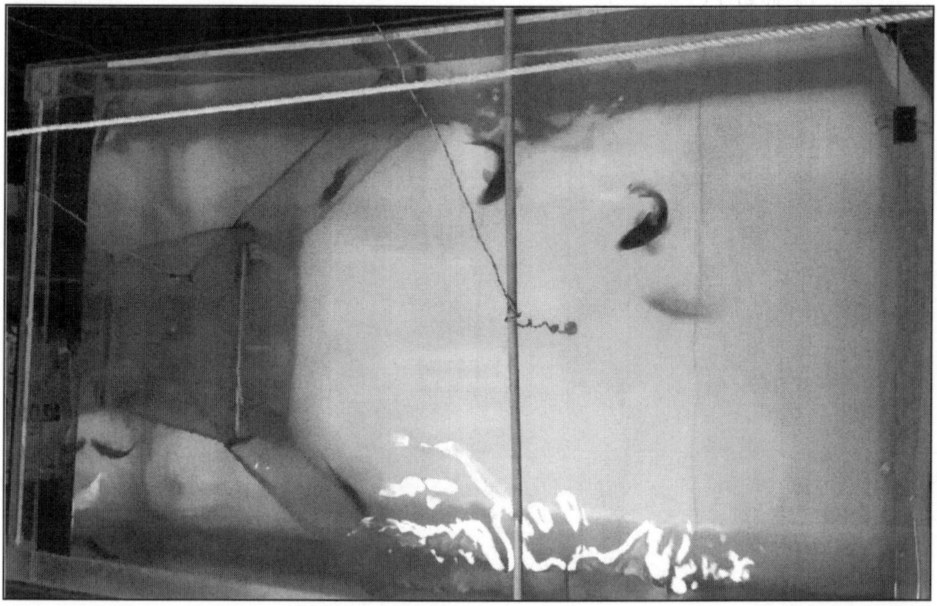

Turning of Fish towards the Positive Electrode.

Experiments were conducted in DC and ODC 3 Hz in waters having an electrical conductivity of 0.3-0.7 micro Siemen and 1,12 micro Siemen. In six sets of experiments with DC field, at an input voltages of 120-130 V, 47-58 per cent of mixed fish population moved towards the anode but, 11-28 per cent of them could enter into the trap net behind the positive electrode.

Species like, *C. mrigala, N. notopterus* and *Cyprinus carpio* were found to jump towards the positive electrode. Instant movement on switching on the current were exhibited by the fishes of these species, but could not reach to the anode, dropped at the bottom, lied on its sides or belly up, ceasing opercular beat as long as the current flows. On switching off the current, fishes were found to recover from immobility and started swimming as before. It is presumed that at the instant current flow at a pre-set voltage, fish after the first leap towards the anode was unable to exercise any muscular movement, since the pre-set input voltage exceeded the threshold values of taxis, which could have been avoided if the voltage would have been increased very slowly. But this was practically impossible to observe the reactions of so many fishes of different sizes at a time.

The alternative was to use oscillating direct current (ODC). The same fish population, when subjected to ODC 3Hz, only 65-80 V were necessary to influence 51-63 per cent fishes to swim forcefully towards the positive electrode and 24-39 per cent to enter in the trap net. Even after switching off the current flow 39 per cent of the total fish population (*C. punctatus*-9 per cent, *L. rohita*-6 per cent, *C.carpio*- 6 per cent, *L. bata* -6 per cent, *C. mrigala* -6 per cent, *O. niloticus* -3 per cent, *H. fossilis* -3 per cent) was found to remain within the trap net, not came out of the net even if they were physically disturbed.

Changing the size of the trap net of 47 X 36 cm with 46 X36 cm wings and the configuration of anode in the large tank (7 droppers, 3-1-3 pattern, leaving 16 cm central gap), 100 per cent anodic taxis and entry into the trap net was observed in case of *L. rohita, L. bata and C. carpio*, when they were exposed for 18-72 seconds in a voltage of 32-42 V in ODC 3 Hz. The current density registered as 140-220 micro-amp/cm sq. Other species like, *C. punctatus, C. striatus* exhibited 80-100 per cent anodic movement and entry into the trap net at 30-50 volts (current density being 220-240 micro-amp./cm.sq.) within 57-95 seconds. The current type was ODC 3 Hz.

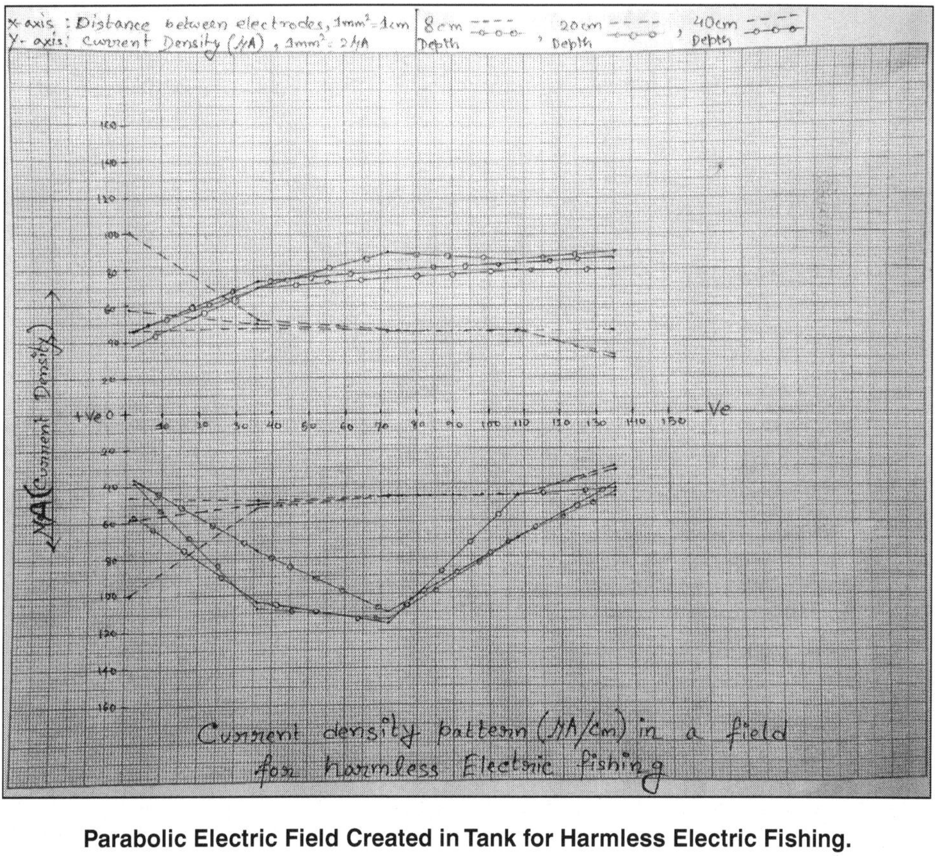

Parabolic Electric Field Created in Tank for Harmless Electric Fishing.

Claims for the Invention

While studying the behavior of nine freshwater fish species, common in Indian waters, in an electric field, it has been observed that after perception of electric field, fishes seek a safe area, where there will be no influence of electric field. This was evident when the fish, under the influence of higher current intensity, slightly above perception level, violently moved between the electrodes, jumped out of the field and tried to find out a space outside the effective zone of the electric field.

Attempts have been made to quantify this reaction of avoidance in an electric field by using a trap net behind the positive electrode (an aluminum ring of 14.3 cm diameter on which 8 aluminum droppers of 20 cm length were suspended on copper wire stirrup, so that solid aluminum droppers hang loosely and can swing freely without disconnecting electrically). Fishes seeking a no influence zone after perception of electric field and on reaching the threshold value of galvanotaxis were found to swim forcefully towards the anode (ring with droppers) set aside the swinging droppers and entered into the trap net behind the anode

Care was taken to regulate the intensity of electric field, so as not to succumb the fish for narcosis or titany. For this purpose, the current intensity of the field was raised very slowly, so as to reach the exact threshold intensities to initiate the rapid swimming of fish towards the positive electrode.

Fishes of each species were tested individually in order to determine the exact value of threshold intensities for involuntary forced swimming towards the positive electrode with respect to size (length) of fishes, and electrical conductivity of water. There after fishes in groups and multiple species in groups were subjected to electric exposure and the percentage of fish movement towards the anode and entry into the trap net were determined at minimum input voltages

None of the fishes of any species was found to suffer any injury, as all of them at the later stage of perception or at the beginning of galvanotaxis swam to the anode, forcefully, passed the droppers and took shelter in the trap net behind the droppers. The success of this new technique lied in using low input voltages and determining the threshold values of these reactions for each species.

A voltage requirement of 32-86 V, 50-114 V, 22-32 V and 32-60 V for 100 per cent capture of *L. rohita* fingerlings (130-199 mm) in DC, ODC 1Hz, ODC 3 Hz and ODC 6 Hz was determined in narrow fields (small tank) trials. In a wider field (big tank), there was marginal increase in voltage requirements (42-100 V in DC; 80-170 V in ODC 1 Hz and 45-62 V in ODC 3 Hz) for bringing out 100 per cent galvanotactic effect and entry into the trap net. As regards the hundred percent desirable response in the lower voltage range, ODC was found to be superior than DC. But the requirement of higher threshold voltage in ODC 1 Hz, than DC was due to slowest pulse rate and shortest effective period. The fish after receiving first stimulation got enough time to react for fright reflex in the same voltage. This was checked by increasing the voltage than of previous stimulation, which forced the animal to swim towards the anode and to the trap net behind. The same trend were followed by the other carp (*L. bata*) and scale less catfish (*C. batrachus*).

The exception was in case of Nile tilapia (*O. niloticus*) when only 33-66 per cent of them responded for galvanotaxis and capture in DC of 120-130 V. By increasing the voltage to 160-220 V, hundred percent of them could be captured in the trap net in ODC 1 Hz.

Freshwater prawn (*M. rosenbergii*) did not respond for the galvanotactic reaction in DC even at 40-100 V. Fifty percent of them in 70-120 V and 100 per cent of them in 18-30 V moved to the anode and in the net in ODC 1 Hz and ODC 3 Hz respectively in the small tank. In wider area (big tank), only in ODC 3 Hz, hundred percent of them responded for galvanotaxis and capture

In multi-species fish population in wider field (big tank), 47-58 per cent could be assembled near positive electrode in DC field at an input voltage of 120-130 V, of which 28 per cent entered in the trap net. The result was improved in ODC 3 Hz, when in much lower input voltage of 65-80 V, 51-63 per cent of total fish population moved to positive electrode and 33 per cent of them entered and remained in the trap net even after switching off the current flow.

By changing over to a comparatively larger trap net (47 X 36 cm opening) and with different anodic arrangement (seven 21 cm long aluminum tube droppers, hanging loosely, in front of trap net opening, one at the center and three on each side of the net opening), 80-100 per cent of mixed fish population swam forcefully to the positive electrode and in the trap net, when they were treated in ODC 3 Hz.

Summing up it can be concluded that, this new technique could catch fish in a trap net with low voltage electric stimulation without causing any harm or injury to fishes. Of three types of current studied for stimulating fish to fright reflex, attraction to the anode and entrapment ODC was found superior to DC. Among three frequencies tested, ODC 3 Hz, followed by ODC 6 Hz were found still superior to stimulate fish and prawn for taxis and capture in the trap net. As regards the specific sensitivity, freshwater prawn (*M. rosenbergii*) was most susceptible in lower voltage of 18-30 V followed by major carp (*L. rohita*) in 22-32 V, scale less catfish (*C. batrachus*) in 25-42 V, minor carp (*L. bata*) in 22-60 V and Nile tilapia (*O. niloticus*) in 26-65 V.

Sharber *et a*l. (1994) observed that spinal injuries in fish in an electric field are associated with higher frequencies and can occur at any point after the onset of galvanotaxis. In the present trials low frequencies (1 Hz to 6 Hz) was used and care was taken not to increase the input voltage beyond the threshold values of galvanotaxis, which had brought success to the technique of harmless electric stimulation for fish capture.

Justification for the Claim

Comparative study of some of the prior arts related to the new technique for harmless electric fishing

Pub. No. CN 201248322Y

Ship borne electric fish catching, electrically hauling net component, where two or multiple boats are used for guiding the drawing mesh. The machine automatically traps the fish into the mesh, guides and draws the electric mesh back and forth for almost catching the fishes contacting the mesh. The upper end of the drawing mesh is connected with the anodic output and the lower end the lower end is connected with cathodic output end of the converter. It was claimed that the device was used for catching fishes avoiding water drainage, water pumping and drying pond and saves labor and time providing convenience for fishermen.

In the publication CN 201248322Y, mention has not been made about the drawing of electric mesh back and forth and the material with which the electric mesh was made of. How the upper end of the drawing mesh was connected to the anodic output and lower end was connected to the cathodic output without being short circuited, if the drawing mesh was of metallic origin? No mention was made about the strength of the electric parameters (voltage between the anode and cathode) applied in the drawing mesh and the drainage of current (ampere) during operation. It was also stated that fishes were caught contacting the electric mesh, when drawn back and forth, which means injuries and death of fish coming in contact with electric mesh.

The new technique for harmless electric fishing was entirely different from the above cited publication. To avoid injuries and mortality of fish during electric fishing with higher current intensities, like, bent spine, fracture of vertebrae, internal hemorrhage and burns of body in direct contact with the electrodes, this new technique has been devised to create sub-lethal Polaroid electric field.

Electrodes have been specially designed and shaped with the cathode 2-5 times larger than the anode in dimension, and arranged in such a manner, so as to obtain Polaroid electric field with a depression channel at the center of the anode (anodic droppers) where there was little influence of electric field v(perception zone). Fishes from wider area after electrically stimulated (at sub-lethal thresholds) were forced to swim towards the anode (galvanotactic effect) and were found to search frantically an escape way out of the electric field. On its endeavour to search for an escape near the positive electrode fishes would found the depression channel and immediately took refuge in the trap net through the depression channel.

The electrical stimulation of fish for involuntary swimming towards the anode never exceeded the threshold value of galvanotactic level and the effective period of current flow (exposure time) did not exceed 130 seconds. Consequently there was no chance of any harm or injuries of fish caught under this technique.

As such, this new technique cannot be compared with the above mentioned publication.

Pub. No. CN 105472855

Electronic remote pulse-type fishing device, described only one component (pulser), that is, the circuit by which pulse type of direct current can be created and applied in electric fishing. It related to the power supply to electric fishing when pulsed direct current was used. It was only a part of the electric fishing system, and not the whole system of electric fishing.

The new technique for harmless electric fishing, therefore, cannot be compared with the above mentioned publication.

Pub. No. CN 204297042

Electric drive long line fishing boat with photovoltaic system related to the electric propulsion system of a long line fishing boat and is not connected with the fishing device. As such, the invention was not in any way comparable to the new technique of harmless electric fishing.

Pub. No. JP 10248465

Electric shock imparting device for fishing. The device was stated for use with a fishing tackle having a hook attached to a tip of fishing line, comprising a DC power supply. It appears that the inventor intended to use the device in line fishing (may be hook and line or pole and line or long line, which is not clear from his presentation) to shock large sized fish which are very strong and give a good fight when hooked. From the drawing given in the publication, it appears that the device was meant for line fishing (fishing with hooks and line) in the sea.

This device is no way comparable to the invention of new technique for harmless electric fishing.

Pub. No. CN 205547047

Fishing cage is trapped in spontaneously electro-luminescence – The self generating luminous trap fishing cage is utilized for luring the fish into the trap to be caught.

This device was based on the phototaxis behavior of the fish to entrap them for capture. Only the fish with positive phototaxis behavior (attractive to light) will be attracted and enter into the fishing cage, which is entirely a different approach than to that of the new technique for harmless electric stimulation to attract the fish towards the anode and into the trap net through the depression channel in frenzy of finding an escape from sub-lethal electric field, thus set up between the anode and cathode.

Pub. No. CN 2355535

Electric net for tucking fish, which relates to an electric net mounted on a frame and a floater to draw fish together into a small space with the help of pulse power source. It was a fish gathering device and not the fish catching device. With the help of pulse electric current, fishes were gathered in a small area, so that fishermen can catch fish with net or other fishing gears. The new technique for harmless electric fishing is completely different from the device mentioned in the above publication.

Pub. No. US 4839986(A)

Electric fish landing net prevents fish from struggling and escape, when they are concentrated into the brailer to lift fish out of water for landing on the fishing boat. The device in other form is used in purse seine fishery, when tonnes of fish are pursed and concentrated within short space of the bunt portion of the purse seine net. To prevent jumping and struggling of fish within small pocket of the net, fishes were stunned with such device to arrest quick spoilage due to lactic acid formation within fish body by struggling in a small space. The new technique of harmless electric fishing is no way comparable to this device under above publication.

Pub. No. US 20100200308

Fishing net incorporating electronic weighing device to weigh fish whilst the fish is still in the net. The invention was completely different from harmless electric fishing, where fish was caught with sub-lethal electric stimulation without any harm or injury.

Pub. No. JP 2013215117 (A)

Tuna farming, using fish farming net, to which alternating or pulsed current is passed through net webbing by weaving a conductive thread in the preserve net. The device is an electrically protective wall inside the tuna fish reserve to prevent tuna from colliding with the fish preserve nets (nets used for tuna ranching). Otherwise tuna will collide with the net and die from neck fracture. The device of using AC or pulsed current as a protective wall for tuna ranching is no way comparable to the new technique for harmless fishing, where fishes were caught by attracting them to the anode first and provided an outlet to escape in the trap net.

Comparing with previous publications, in the area, and summing up the differences identified in the present invention are:

1. To mitigate the problems of harmful injuries and death of fish during electric fishing, this new technique of harmless electric fishing was conceived, designed and tried with success.
2. Electrodes, especially, positive electrodes were specially designed and arranged in such a way, so that a Polaroid electric field could be created with a depression channel near the anode.
3. Fishes in a sub-lethal Polaroid field, being electrically stimulated for taxis towards the positive electrode, swam involuntarily and tried to escape out of the electric field on reaching the anode.
4. On finding a depression channel(where a feeble electric field persisted), fishes escaped out of the electric field through the depression channel and entered into the trap net behind and caught.

Problem Solved by the Invention

The problem of harmful effects, such as, injuries of fish (both external and internal injuries) and immediate and delayed mortality of fish during and after electro-fishing could be prevented by adopting this new technique of harmless electric fishing.

Solution Provided by the Invention

In the invention of new technique for harmless electric fishing, fishes were attracted towards the anode (positive electrode) by creating a sub-lethal electric field with the help of electrodes of proper materials, size and shape, arranging them in such a way that a Polaroid field was created between the electrodes with a depression tunnel, where the field intensity did not exceed the perception level. Fishes after exhibiting forced swimming under the influence of sub-lethal electric field, towards the positive electrode and frantically searching for an escape out of the sub-lethal field, which the fish felt strongly on reaching the positive electrode, found the depression tunnel and moved towards the tunnel, where they could perceive electric field intensity at a perception level, moved forward and entered into the trap net, where no electric field existed. It was found that fishes congregated one after another in the trap net and tried to remain in the trap net, even after stoppage of current flow. When, after stoppage of current flow, fishes were removed, physically, they were reluctant to come out of the trap net. No immediate or delayed mortalities were noticed due to external or internal injuries of fish by this method of fish catching.

Advantages

The main advantage was to catch a large number of fish (depending on the capacity of trap net) with electric stimulation, unharmed, taking the advantage of galvanotactic reaction of fish (forced involuntary swimming towards the positive electrode) and escape of fish near the anode (positive electrode) through a depression channel, created by suitable electrode arrangements, (low field intensity at a perception level). Fishes, on receiving a sub-lethal electric stimulation, in a frenzy of fright reflex, rushed towards the depression channel and into the trap net behind the positive electrode.

There will be no immediate or delayed mortality by this new technique, which normally happened, to some extent, during electric fishing.

By this technique fishes can be caught from stagnant waters, like, ponds, tanks and reservoirs.

4

Design of a New Three-Walled Electric Gill Net to Attract and Catch Fresh Water Fish

Conceived, Designed and Trial – Dr. K.P. Biswas

Associate in the Work – Mr. Subhasis Dolui

Background of Invention

Gill nets are used all over the world to catch every kind of fish, because the structure of the gear is very simple and one of the cheapest fishing gear. Gill net consists of a rectangular net with float on its top and sinkers on its bottom, in order to spread the net in the water. The gill net is placed in the way of the swimming fish. Attempting to pass through the net fish are enmeshed or entangled in the net. Generally spindle shaped fish, such as, sardine, mackerel, salmon, yellowtail and skipjack are enmeshed; while crab, prawn, spiny lobster, flatfish, shark, Spanish mackerel are entangled. From its catching frequency, the gill net fishing gear has two different sorts of functions, enmeshing and entangling. In designing the modern gill net, therefore, the following matters are to be taken into consideration, (a) how to increase the catching frequency and (b) fish species selectivity of the fishing gear.

While the mesh size is a crucial factor in catching the fish, it also selects the fish caught by body size or shape. The gill net is by far the most common place for the fish to be caught in the net. The part of the net in which the most fish are caught is the lower fifty meshes of the lower net. Although it is not really a remarkable amount, the greater part of the catch seems to be caught in the central portions of the net rather than the ends. And very few fish appear to be caught near the seams, where the net has been sewn together.

Therefore, in designing a gill net fishing gear, one must select a type that would serve particular purpose best.

The inland fishery resources of India comprise of rivers and canals, reservoirs, tanks and ponds, estuaries, brackish water lakes, flood plain lakes *etc.* The gill net is a predominant and popular fishing gear in inland traditional fish harvesting sector, because of its simplicity in design, fabrication and low man-power and energy requirement for operation. It is operating in eighty percent inland water bodies. Gill net can be operated in waters where other dragging gears, such as, trawl and seines are unsuitable due to physical obstructions (Biswas, 1996).

The principles behind gill netting have not changed over the years, but the equipment and materials have changed. It is widely recognized as an efficient and "selective" type of gear (Bjoringsoy, 1996). Catchability and selectivity of gill net is influenced by several factors, like, mesh size, visibility of net material, stretchability of meshes, net construction, method of fishing, shape of fish and its migration and movement pattern.

Visible signs of fish reaction in an electric field attracted a good deal of attention in controlling fish movement for capture and management. The definite sequence of reactions with an increase of field strength was observed. Investigations were carried out with different types of current, various species and length of fish, their physiological condition and the environmental parameters, like temperature and electrical conductivity of water.

Electric current in water dissociates molecule of substances, known as electrolytic dissociation. When the conductivity of water is less than the fish body, as in freshwater, all the potential lines are directed to the fish body with better conductivity and the fishes thus are satisfactorily influenced (Biswas, 2015).

In order to produce a fishing effect at a certain distance from the electrode, current of such strength must flow that a fish of length (L) can receive a body voltage (G) for swimming to the positive electrode. The current density (I) that must exist around the fish to effect for this reaction can be computed from the length of the fish, body voltage (potential difference between head and tail) and the specific resistance of the water in which they inhabit.

According to the varying effect of electric current on the fish, a distinction can be made between the fishing gears;

1. Attracting the fish towards positive electrode (anodic effect);
2. Frightening or blocking effect; and
3. Narcotizing or killing effect.

All these groups are used in freshwater and sea fishing.

For attracting fish in fishing, direct current (DC) and modulated direct current into pulse direct current (PDC) are used. The rate of pulses can be varied according to fish species from 3 to 60 pulses per second.

The practical electric fishing developed mainly in Germany with Holzer (1932) and the support of physiological research of Scheminsky (1924). However, early in 20th century, Japanese fishermen were using induction coil to drive eels out of their burrows into a net (Takahashi, 1895).

The response of Indian fishes were identified in threshold current densities in a homogeneous field (constant current intensity) of large experimental tanks, in which the fishes were subjected to electric exposure (Biswas, 1971).

Temperate fish, like, *Salmo iridius*, could sense the presence of electric field by quivering motion of the body or dorsal fin, which occurs in the perception zone around the electrode. The first reaction does not occur if the DC field is weak or applied slowly. The fish after exhibiting the first reaction moves into the effective zone around the electrode, which correspond to the increased voltage, current density or proximity to the anode, and depend on the orientation of fish in the electric field (parallel or perpendicular to the lines of current conduction) and if the fish is facing anode or cathode.

Gill net (a rectangular net with floats on the top and sinkers on its bottom) stand vertically, like a wall in a water column when spread in the water. Attempting to pass through the net wall fish are enmeshed or entangled in the net. As the gill net does not chase the fish for capture and acts as a passive fishing gear, fishes are gilled or entangled during migratory movement vertically during search for food and horizontally to explore new feeding and breeding areas. Due to this, the catch per unit of effort is low and the fisher has to wait for a reasonably long time before lifting the net and find out the catch.

Mild water current acts as a stimulus for such migratory movement and that is why gill net is operated in near shore areas, subjected to mild water movement. But in case of standing still water, devoid of water current and movement, such as, reservoirs, ponds and tanks, chances of fish being gilled and entangled become remote, except at night, when visibility is poor. In any case, ordinary gill net fishing is a chance fishing and time consuming fishery.

It was therefore; thought if fish could be stimulated to move in a standing water, then there might be a chance for increased enmeshing and entangling in the net. In fact, the conventional drive-in-fishery is based on this principle, where the whole village participate in driving the fish by beating and splashing water towards a barricade of gill net, which stand like a wall in the middle of the water area. Fishes being frightened by the beating and splashing of water move frantically and being gilled and entangled in the gill net spread in the mid-section of the water area.

The Invention

A prototype three-walled electric gill net was designed and fabricated with the objective that, it would have better fish catching efficiency by creating sufficient electrical stimulus around the net through the electrodes.

Three-panel gill net of following dimensions and specifications was fabricated for experimental trials.

1. 0.8 X 0.8 m framed gill net having uniform mesh of 2.5 cm (2 outer panels).
2. 0.8 X 0.8 m framed gill net with uniform mesh of 1.5 cm (central panel).

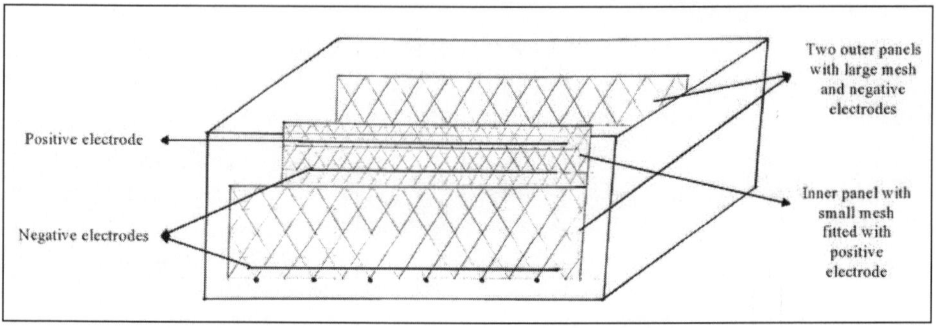

Schematic Diagram of Experimental Tank Showing Electrified Gill Net with Positive and Negative Electrode (Not to the scale).

In the first set of experiments, the net was fitted with 0.8 m 14 gauge bare copper wire, each at the bottom of two outer panels, which served as negative electrodes (cathode). The positive electrode, which was also 0.8 m 14 gauge bare copper wire, was attached to the upper part of the central panel, 4 cm below the float line. During operation, the positive electrode remained under the water surface. The distance between the central panels to the outer panel was maintained 0.42 m at each side.

The negative electrodes at the bottom of two outer panels were electrically connected by 145 cm insulated cable on each side, 25 cm of which was kept un-insulated and exposed in the middle, so that, spike potential could be developed under the positive electrode to form two concentric triangular field, when a potential difference is applied between the anode and cathode. In fact, the mid-exposed cable in both the sides of anode also acted as negative electrode, creating multiple tangential fields through the water column, when the net was energized.

The second set of experimental gill net was fitted with aluminum conduit of 12 mm diameter, except the copper cable connecting the two negative electrodes.

The same dimensioned gill net panels, without electric connection were used as conventional gill net for comparison in the trials.

Experimental trials were carried out for a period of four and half months. A total of 15 sets of trials were conducted during the entire period. The experiments were carried out during day time. Electric stimulation of 2 to 60 seconds were applied to the fishes, and the affected fishes, enmeshed and entangled in electric gill net were removed immediately and kept in highly oxygenated water for a period of 30 minutes, which was maintained throughout the trials.

Measurement of Electrical Parameters during Operation of Electric Gill Net.

A glass tank having hydrodynamic facilities for identification and quantification of observations of submerged objects in stagnant water was used for conducting the trials. The overall dimensions of the tank was 1.8 m long, 1.05 m wide and 0.8 m high. It was filled with freshwater up to a depth of 0.6 m in a temperature range of 25 to 32 degree Celsius, and electrical resistance of 4000 to 6500 ohms/cm.

One 0.5 kVA AC to DC inverter, capable of supplying 40 to 230 volt direct current (DC) was employed as a source of power supply to the gill net. An electronic pulser of low frequency (1/sec to 3/sec) was connected in the outer circuit between inverter input and AC supply line to obtain pulsed direct current (PDC), which was transmitted to the electrodes.

A total of nine species of common Indian freshwater fishes were used for the trials with electric gill net. They were:

1. *Labeo rohita*, under the major carp family with a curved dorsal profile with obtuse snout, projecting slightly beyond the jaws without lateral lobe. A short and thin pair of maxillary barbells with deeply forked caudal fin and bluish or brownish color along the back, becoming silvery on the sides and beneath are the prominent features of the species. Their fins are diffusely banded grey and red and scales with dark margins and red center or sometimes orange buff inside a dark edging. They grows to a length of one meter maximum and inhabit in deep pools and clear sluggish rivers throughout India. The species is largely used to stock ponds for culture.

2. *Cirrhinus mrigala*, another member of Indian major carp family is characterized with a wider mouth and thin lips. Lower jaw has a small tubercle in the center. Small barbels in fold of lip with sharply forked caudal fin have large scales. Their color is dark grey along the back, silvery on the sides and below having a coppery tinge in pectoral and ventral fins. Anal fin is orange in color, stained with black. They grow to a maximum size of nearly one meter and resides in rivers and tanks of India. The species is used in stocking ponds for culture.
3. The body color of *Labeo bata*, belonging to Indian minor carps, varies with the age of fish, generally silvery, darkest along the back and the lower fins stained orange color. Fine black dots are present on all the fins. When about 10 cm long, there are 3 or 4 small black spots on the fifth and sixth scales of the lateral line, which gradually and almost entirely fade as age advances. Sixteen to eighteen centimeter lengths are common and is extensively used for stocking tanks and ponds with major carps.
4. *Notopterus notopterus* with laterally compressed body has convex dorsal and ventral profiles and the whole appearance rendered unusual by the long anal fin which extends for more than two-thirds of the whole length of the fish and is confluent with and almost marks a very small caudal fin. The dorsal fin is short and pelvic fins are very small. The body is silvery in color with minute scales and grayish spots everywhere. They are available in fresh and brackish waters of India and normally attain a size of 25 cm.
5. The body shape of *Oreochromis niloticus* is laterally compressed to oval and deep, depending on the environment. Their body is covered with cycloid scales. It has a terminal mouth. Caudal fin has 7-12 distinct vertical strips. Caudal finis truncate. The most diagnostic features are the regular and definite strips on the caudal fin. The red flush of the breeding male and the dark margin of dorsal fin is conspicuous. Their body color varies depending on the environmental, physiological and dietary factors. They attain a maximum length of 60 cm.
6. Known as common carp, *Cyprinus carpio* has one long dorsal fin, which possesses 2-3 hard and 17-22 soft rays. The color of the body varies from grey through silver to bronze with a yellowish or reddish belly. Their mouth is large with two pairs of barbels, one pair on the upper lip and the other pair at the corner of the mouth. In wild, common carp live in the middle or lower sections of rivers and in areas where the water is shallow and the bottom is muddy, in a temperature range of 3-35 degree Celsius. Common carp, an invasive species of India were introduced in fresh water rivers and reservoirs and are widely cultured in India along with major carps.
7. *Labeo calbasu*, one of the member of Indian major carp family, has a deep stockily-built body. Pores on snout and upper lip, chin slopes away sharply from mouth are prominent features of the species. *Rostral barbels*

are slightly longer than maxillary and about equal in length to diameter of the orbit. Their caudal fin is deeply forked. The body color of the fish is dusky and each scale with a pinkish buff at the center and a dark margin. The margins sometimes form dark longitudinal lines. Fins are dark and iris rich coppery. It attains a length of 50 cm. It is said to be a tank fish, but inhabit in deep pools in clear sluggish streams of India.

8. *Heteropneustis fossilis* is a flat and white headed fish with almost chisel-like snout. The body tapers uniformly in the horizontal plane from the gills to the caudal fin. In the vertical plane, the back is almost a straight line from snout to tail, while the ventral outline is bowed. This effect is heightened by the shape of the long anal fin. They attain a total length of 12 to 20 cm. Maxillary paired barbels reach the pectoral fins or further. The sharp and serrated pectoral spines are poisonous. Dark purplish brown in color, the young ones are reddish brown. There are sometimes two longitudinal yellow bands. Attain to a maximum size of 25 cm, they are found in rivers and tanks of India.
9. The body of *Channa punctatus* is sub-cylindrical, tapering from the flattened snake like head to the rounded caudal fin. Brownish color on the back fading to lighter beneath. A series of vertical darker bands above the lateral line, alternating with a similar series below it. The last band before the caudal fin is continuous above and below the lateral line. Pectoral fins are striated. Generally attain a length of 17 to 18 cm. They are found in the muddy streams and ponds of India. Stagnant water is favored to running water.

Three panel electric gill net, fortified with electrodes (positive electrode in the central panel and negative electrodes in the two outer panels) were spread in the experimental glass tank. A known number of multi-species fish population was introduced in the experimental tank after measuring the water temperature and electrical resistance of the water with the help of a thermometer and ohm-meter. Fishes were allowed to be used with the new environment for 30 minutes.

The same gill net panels without electric connection was put into the tank for half an hour and then lifted to find out, the number and type of fishes gilled or entangled in the net. After removing the gilled and entangled fish alive, and taking their length and weight measurements, they were released back in the experimental tank and waited for another 30 minutes to adjust with the environment. The gill net panels were again placed in the tank centrally (one with positive electrode at the center and the two outer panels, fitted with negative electrodes, one at each side, at a distance of 0.42 meter from the central panel) and was energized. Electric current (DC or PDC) from the inverter and pulser was allowed to pass to the positive and negative electrodes, the voltage of which was measured through a voltmeter connected in the circuit. The density of current was, however, measured with the help of a probe of one square centimeter to directly read the amount of current flow in the field.

Switching on the current to the electrodes, fishes between three net panels were found to swim forcefully either to the central panel containing positive electrode, being attracted to the positive electrode, or entangled in two outer panels while trying to escape out of the field around negative electrodes. Electric exposure lasted for 2 to 60 seconds and the enmeshed and entangled fishes were removed after switching off the current flow, measured for length and weight, species-wise.

Changing the copper wire electrodes with 12 mm diameter aluminum conduits for the anode and cathodes, trials were conducted in a similar manner in order to find out the electrode type to give the highest catch between two different types of electrodes

The voltage gradient of the electric field, generated between the electrodes of the net panels was measured with a millivoltmeter and the current density of the mid-field was determined with a probe electrodes of one square centimeter at different input voltages to understand the electric field pattern to which fishes were subjected. Fishes of different species combination, size groups were experimented, besides fishes of same species individually, duo and trio for detailed observations.

All the electrical parameters during experiments were measured with two precision meters (AVO, model-8 and Sanwa P-2) and electrical resistance of water and water temperature were recorded with the help of ohm meter and thermometer respectively.

Independent sample "t" test was carried out to compare the effectiveness of two types of electrodes. The relationship between the number of fish caught and effectiveness of two electrode types were analyzed using a linear regression model. The SPSS (version 22) statistical package was used for all statistical analysis.

Significant Findings

Since in three panel electric gill net, the anode and cathodes were attached in such a manner (anode in the central panel in mid-section, just immersed under the water surface and the cathodes at the bottom of the outer panels at the tank extremities, interconnected by insulated cables in both sides, keeping a central 25 cm length of the cable without insulation, exposed), the electric field that was created between the electrodes was of heterogeneous nature. Therefore the voltage gradient and current density pattern of the electric field between positive and negative electrodes were not uniform. Their strength varied at different points between the electrodes.

It was found that the current density progressively increased from 0.005 mA/sq.cm to 7.5 mA/sq. cm, when the input voltage increased from 10 V to 175 V, when horizontal aluminum conduits were used as positive and negative electrodes. A two-dimensional current density at tangential distances between negative and positive electrodes was measured at the same electrode configuration, when the input voltage was 85 volts.

Highest current density of 40 mA/sq.cm was observed at 77 cm away from negative electrode and 6 cm away from the positive electrode. The lowest current density of 3.5 mA/sq.cm was recorded at 28 cm and 59 cm away from cathode and anode respectively.

The voltage gradient of the field at two dimensions, that is, tangential and vertical distance from the negative electrode (cathode), at an input voltage of 85 V revealed that a voltage gradient of 10 V/cm at a tangential distance of 10 cm from the cathode, gradually increased up to35 V/cm, when the tangential distance increased to 80 cm from the negative electrode. The voltage gradient at 5 cm (vertical distance) from the negative electrode was 15 V/cm (nearer to the positive electrode). At 35 cm vertically from the negative electrode, the voltage gradient was found to be 18.5 V/cm.

Three dimensional voltage gradient at an input voltage of 90 V, revealed, that the horizontal distances from the anode and cathode together, with the vertical distance between them gradually increased from 6.5 V/cm to 32 V/cm when the distance from the negative electrode increased from 10 to 80 cm. Conversely, the voltage gradient from 10 cm away from positive electrode (34 V/cm) progressively decreased to 10 V/cm, when the distance increased to 80 cm. In case of vertical distance between the anode and cathode, however, the changes in voltage gradient over the depth (vertical distance) was not so wide. At a depth of 5 cm, the voltage gradient was 18 V/cm, while at a depth of 44 cm, the voltage gradient was found to be 13 V/cm.

There was no fish catch by gilling or entangling in the non-electrified net even after waiting for one hour. Sixteen to eighty four percent of total fish population of the experimental tank were enmeshed and entangled, when the net panels were energized with 40-250 volts, producing a current density of 0.03 to 1.25 mA/sq.cm in the mid-field depending on the intensity of applied voltage to the electrodes. The current exposure was 29-90 seconds, when the water temperature and water resistivity were 25 to 25.5 degree Celsius and 6000 to 6500 ohm/sq.cm respectively.

In a slightly higher temperature of 28-29 degree Celsius and lower resistance of 5200- 5700 ohm/sq. cm, 52-96 per cent of total fish population were caught in electrified gill net, when 95-210 volts were applied between the electrodes, creating a current density of 0.07-0.25 mA/sq.cm.

Changing the material and shape of electrodes with 12 mm diameter aluminum conduit, the electrified gill net could catch within 10 to 30 seconds, 40-83 per cent of total fish population by entangling and gilling, when a potential difference of 65-145 V was applied between positive and negative electrodes, producing a current density of 0.055- 0.25 mA/sq. cm in waters having an electrical resistance of 5200-5500 ohm/sq.cm at 29-30 degree Celsius.

The non-electrified gill net could not catch any fish.

Experiments with DC (direct current) and PDC (pulsed direct current) with 30-110 volts at the electrodes established a current density of 0.01-0.10 mA/sq.cm in the effective zone could enmesh and entangled and caught a good number of fishes within 5-80 seconds current exposure. Highest catch of 75 per cent (53-75 per cent) was observed in PDC-3/sec, followed by DC (74 per cent), PDC-1/sec (53-66 per cent) and lowest in PDC-2/sec (56 per cent).

Fishes Gilled and Entangled in Electrified Gill Net.

As regards the response of different fish species, 100 per cent of six varieties of fish species (*L. rohita, C. mrigala, L. bata, O, niloticus, C. punctatus and H. fossilis)* were caught by entangling in the electrified gill net, where a current density of 0.045-0.500 mA/sq. cm was established with aluminum conduit electrodes.

Gilling of Fish around the Positive Electrode.

Hundred percent anodic attraction and capture of *L. rohita* in electrified gill net was noticed, when they were 52-70 cm away from the positive electrode at a current density of 0.075 mA/sq. cm. At increased current density of 0.9 mA/sq.cm, 100 per cent anodic attraction and capture of the species was observed from 58 and 45 cm away from positive electrode.

In *O. niloticus*, cent percent attraction and capture was effected when the fish was 15-17 cm away from the positive electrode at a current density of 0.15 mA/sq.cm. Further away from the positive electrode, namely, 18 cm and 58 cm away from the anode 80 per cent and 50 per cent of fish was captured respectively. Twenty percent and fifty percent fish escaped out of the field.

Out of total fish population gilled or entangled in electrified gill net, the share of each species in a multi-species fish population of the experimental tank was determined in two electrode sets separately.

In the copper wire electrodes, 63-90 per cent of total fish population were gilled or entangled, out of which, *L. rohita* contributed 57-68 per cent, *O. niloticus* 5-18 per cent, *C. punctatus* 8-11 per cent, *H. fossilis* 8-11 per cent and *N.notopterus* 3-5 per cent.

In aluminum conduit electrodes, 71-95 per cent fish were gilled and entangled. Out of them, *L. rohita, C. mrigala, L. bata, H. fossilis, and N. notopterus* contributed 44-67 per cent, 12-31 per cent, 6-25 per cent, 8-12 per cent and 3 per cent respectively.

The percent of excess catch in electrified net over non-electrified conventional net was found to be 94 per cent in *N. notopterus*, 77 per cent in *C. punctatus*, 76 per cent in *L. bata*, 71 per cent in *C. mrigala*, 70 per cent in *L. rohita*, 67 per cent in *C. carpio*, 58 per cent in both *L. calbasu* and *H. fossilis* and 49 per cent in *O. niloticus* when copper wire electrodes were used in a resistive water of 4000-6500 ohm/sq.cm at a temperature between 25-32 degree Celsius.

In aluminum conduit electrodes, *O. niloticus* and *H. fossilis* contributed 100 per cent capture in electrified net, followed by *L. bata* (76 per cent), *C. mrigala* (71 per cent), *L. rohita* (69 per cent), *C. carpio* and *N. notopterus* (67 per cent each) and *L. calbasu* (58 per cent).

In the copper wire electrodes developing a current density of 0.075-0.35 mA/sq.cm 100 per cent *L. rohita* were gilled and entangled in the size range of 121-130 mm onwards. All *O. niloticus* of 91-100 mm was entangled in electrified net; while all *C. punctatus* of 171-180 mm onwards and *H. fossilis* of 141-150 mm onwards were caught in electrified net. In aluminum conduit electrodes, producing a current density of 0.075-0.35 mA/sq.cm, all *L. rohita* of 201-210 mm; *C. mrigala* of 191-220 mm; L. bata of 161-180 mm and *O. niloticus* of 131-140 mm were entangled and caught in electrified gill net.

Irrespective of electrodes type in DC field the percentage of excess catch in electrified net in order of rising current densities, *C. carpio* and *N. notopterus* were found to be caught in 100 per cent cases in current densities of 0.025-0.64 mA/sq.cm followed by 100 per cent *L. bata* in 0.05-0.07 mA/sq.cm; 100 per cent *L. calbasu* and *C. punctatus* in 0.09-0.15 mA/sq.cm; 100 per cent *C. mrigala* in 0.12-0.18 mA/sq.cm; 100 per cent *O. niloticus* in 0.2-0.23 mA/sq.cm. A maximum of 85 per cent *L. rohita* and 91 per cent *H. fossilis* were caught in current densities between 0.09-0.15 mA/sq.cm.

The movement of individual fish towards the positive electrode (central panel) of the electrified net indicated that 100 per cent *L. rohita* escaped when the current density was up to 0.045 mA/sq.cm. Above that current density (0.046-0.12 mA/sq.cm), no fish escaped and 100 per cent of them were attracted to the positive electrode and either gilled or entangled.

Hundred percent *C. mrigala* swam to the positive electrode at current densities of 0.046-0.30 mA/sq.cm.

Hundred percent *O. niloticus* escaped when the current density was in the range of 0.086-0.09 mA/sq.cm. But when the current density was raised to more than 0.146 mA/sq.cm, 100 per cent of them were attracted and moved towards the anode. However, in current densities of 0.176-0.18 mA/sq.cm and 0.346-0.350 mA/sq.cm, only 67 per cent and 50 per cent fishes moved to the positive electrode.

C. punctatus, however, forced to move towards the positive electrode and caught in current densities of 0.121-0.18 mA/sq.cm.

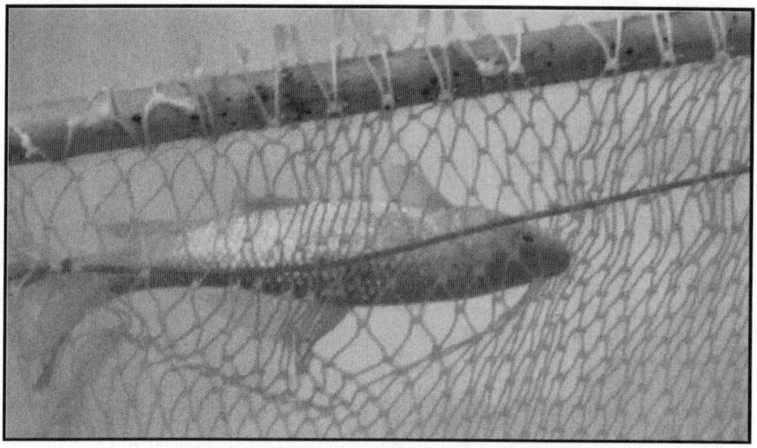

C. mrigala **Entangled around the Positive Electrode of the Central Panel of Electrified Gill Net.**

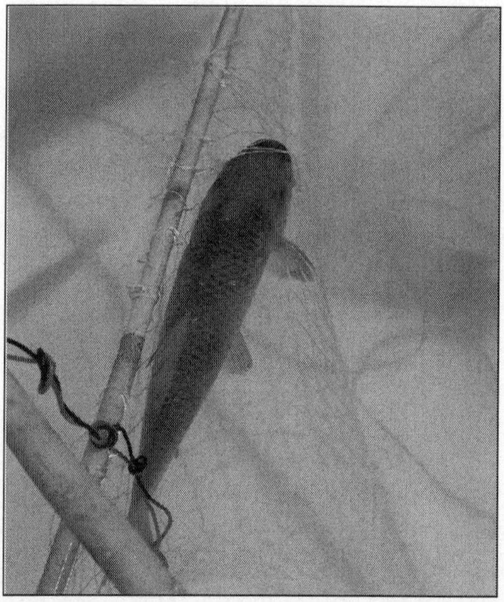

L. rohita **Gilled in the Outer Panel Containing Negative Electrode.**

In a group of 3 fishes of same species, at a time, hundred percent *L. rohita* of 128-236 mm length exhibited forced swimming to the positive electrode in current densities of 0.016-0.5 mA/sq.cm. In lower current densities of 0.011-0.015 mA/sq.cm, only 67 per cent L. rohita was attracted to the anode. Hundred percent *C. mrigala, H. fossilis* and *C. punctatus* showed anodic movement in current densities of 0.216-0.32 mA/sq.cm, 0.011-0.2 mA/sq.cm and 0.016-0.15 mA/sq.cm respectively. 100 per cent, 83 per cent, 33-75 per cent and 33 per cent *O. niloticus* exhibited forced movement and caught in the electric gill net at current densities of 0.096-0.5 mA/sq.cm, 0.746-0.75 mA/sq.cm, 0.046-0.0p8 mA/sq.cm and 0.396-0.4 mA/sq.cm respectively.

Efficiency of the newly designed and fabricated electrified gill net was judged after carrying out independent simple "t" test. Tests were also carried out to compare the effectiveness of two electrode types used in the trials. The relationship between captured fish and the effectiveness of two electrode types in inducing fish to be gilled and entangled was analyzed by linear regression model.

The results indicated that there is no significant difference of fish catch between two electrode systems (t (33) = -0.948, p = 0.350)

The results of linear regression model showed that the total catch in aluminum conduit electrodes was better than the total catch with copper wire electrodes.

Electrode	*Copper Wire Electrodes*	*Aluminum Conduit Electrodes*
N	18	17
Mean	75.44	79.94118
Std. Deviation	14.07	13. 99241
Std. Error (mean)	3.32	3.393657

Claim for the Invention

Since the cylindrical electrodes (14 gauge copper wire and 12 mm diameter aluminum conduit) used in the trials to electrify the gill net, the resistance will depend upon the ratio of separation (distance between electrodes) to the radius of cylindrical electrodes upon the length.

As the radius of the electrodes used were small (6 mm in case of aluminum conduit and 1.5 mm in case of copper wire), the effective zone was small and the increased length of electrode (70 cm) extended the horizontal field further into the water (Cowx and Lamarque, 1990). Increasing radius of the electrodes to increase the effective zone was not possible because of the attachment of the electrodes with the gill net panel and higher weight due to increase of the radius, will cause the net to sag and will substantially decrease the catching capability of the net, when used in conventional way without electrifying the net. Each electrode, surrounded by a region of steady decreasing current density was observed while measuring the field pattern in the water generated by the electrodes.

The sensitivity and reaction of fish possessing a neuro-muscular system depend partly on the characteristics of the applied current. When a steady direct current (DC) of low intensity is passed through water containing fish, animals were frightened.

At higher current densities, the fish tend to swim involuntarily towards the positive electrode. At still higher current densities, the fish will roll over to one side and are incapable of movement (Parker and van Heusen, 1917; Denzer, 1949; Meyer-Waarden, 1957; Halsband, 1959; Lamarque, 1963; Rommel et.al. and Biswas, 1976).

In all the trials of electric gill net with different types of electrode systems, fishes were found to be gilled and entangled in electrified gill net more in numbers, when the current intensity was increased with some variations. The conventional non-electrified gill net failed to catch any fish. Fishes were found to simply avoid the conventional gill net.

As the objective of the trials was to utilize the involuntary swimming of fish towards the positive electrode and to take advantage of the fright reflex of fish around the negative electrode in gilling and entangling them in electrified net, more than 50 per cent (52-96 per cent) catch by enmeshing and entangling in 87 per cent trials (19 out of 22 trials) can be considered nearer to the success.

Earlier works indicated that the reactions, most favorable for electric fishing were observed when pulsating DC was used (Norman and Loukshkin, 1954; Meyer Waarden, 1955 and 1957 and Miyake *et al.*, 1957).

Comparing the suitability among DC and PDC of low frequencies 1/sec. to 3/sec. with regard to enmeshing and entangling of fish in electrified gill net, PDC 3/sec. was found to be superior than PDC 1/sec. and PDC 2/sec. and even from DC to some extent.

According to Halsband (1954), different species of fish react differently to varying pulse rates and as the threshold values of the pulse rates are known, it is possible to select fish to be caught within a certain range of electricity according to size and after determining the threshold current densities for different reactions of fish.

Biswas (1971) observed that the threshold values of current densities initiating forced movement of fish to the positive electrode were different for different fish species. Not only the threshold values of reactions differ from one species to the other, but also differ on the length of the fish of the same species. Larger fish require smaller current densities for the same reaction. This indicates that the threshold values of electrical parameters (current density and voltage gradient) are inversely proportional to the length of fish. A larger fish intercept a greater potential difference between head and tail (body voltage) than smaller ones.

Since the present investigation was capture based, practical electric fishing, and the determination of threshold values for the different reactions was beyond the scope of the work, only the end reaction of fish (forced movement towards electrified net or escape out of the electric field was taken into consideration. Even this investigation indicated that specific differences among fish species did exist. *O. niloticus* required higher values of current density (0.2-0.25 mA/sq.cm) for 100 per cent attraction to electrified net for capture.

The stimulating effect on different size group of same species could not be absolutely significant in the present set up of the trials, due to non-availability of all the size groups of fish species. But then, the trend of their reactions towards the electrified gill net indicated that in a fixed range of current densities (0.075 – 0.35 mA/sq.cm), the percentage of excess catch in electrified gill net gradually increased as the length of fish population increased (45 per cent in 91-100 mm size range to 100 per cent in 161-170 mm size range).

Dolan and Miranda (2003) reported that fish size has been frequently associated with variation in electro-fishing effectiveness. They have observed that the amount of transferred power needed to immobilize 95 per cent of the fish treated ranged from as high as 88.635 micro watt per cubic centimeter darters exposed to PDC 15 to as low as 28 micro watt per cubic centimeter for large bodied fish of several species. They have indexed fish size by total body length, surface area, volume and weight.

The percentage of excess catch in both the electrode combinations (copper wire and aluminum conduit) experimented, indicated that aluminum conduit, as electrodes could create a larger effective zone compared to the copper wire electrodes. As a result, all varieties of fish were attracted and entangled or enmeshed in the electrified gill net with aluminum conduit electrodes. Cent per cent *O. niloticus* and *H. fossilis* were captured in electrified gill net. Fishes like murrels, catfish and featherbacks (*C. punctatus, H. fossilis* and *N. notopterus*) were found to be most sensitive towards the electrified gill net followed by carps (*L. rohita, C. mrigala* and *L. bata*) indicating specific variations.

Dolan and Miranda (2004) reported that large mouth bass was most vulnerable to electro-fishing, while the channel cat fish was least vulnerable and bluegills exhibited effects that were intermediate.

Price and Peterson (2010) observed that electro-fishing capture efficiencies were generally low and differed among species groups based on family or genus.

While experimenting with individual fish or group of duo and trio of same species, it was found that at sub-threshold current densities, species like, *L. rohita* and *O. niloticus* escaped from the effective field and not all of them were attracted to the electrified gill net. This reaction was more erratic in case of *O. niloticus*, which can be explained for the specific resistance of the species. The orientation of the body of *L. rohita* in the electric field with respect to direction of current lines, at the time when they entered in the electric field, was the cause of escape of some fishes from the effective field.

The performance of non-electrified and electrified gill net clearly indicated the superiority of electrified gill net over non-electrified conventional net as there was very little catch in conventional net in all the fishing trials.

While advocating for capture of large bodied fishes by electro-fishing in comparison with fyke net, where only small bodied fishes were caught, Ruetz *et al.* (2007) informed that the species composition of captured fish differed considerably between fyke netting and electro-fishing.

Fisheries scientists have long studied the behavior of fish in electrified water. However, the science of electro-fishing (the use of electricity to catch fish) evolved independently of other fields of science, whose knowledge could have explained the behavior of fish to electric shocks. The classical stimulus-response (S – R) theory was dominated in the late 19th century. The local action theory of electric energy on nerves and muscle-fibers in one form and other remain the main explanation for the galvanotropisms today. The neurologists have recognized for more than hundred years, that electric stimulation of vertebrates causes epileptic seizures. Spike-wave patterns of neural discharge of electroencephalograms, which are diagnostic of epilepsy, have been recorded in fish.

Sharber and Black (1999) suggested that the current electro-fishing terms can be matched with epileptic terminology, twitching, orientation, taxis and the turn or escape maneuver as automation, petit mal seizere and grand mal seizere.

Although no significant difference of effected fish catches between copper wire electrodes and aluminum conduit electrodes was shown in simple "t" test, the linear regression model's results indicated that the total catch affected in aluminum conduit electrodes was better than the total catch affected in copper wire electrodes.

Summing up the results of the trials, the electrified gill net was found undoubtedly superior in gilling and entangling of all the Indian freshwater fish species taken up for the trials, when the non-electrified gill net failed to catch any fish.

Among DC and PDC, PDC 3/sec was found to be most effective for driving fishes to the electrified gill net for capture, followed by PDC 1/sec and PDC 2/sec.

Fishes of different species responded variedly, murrels, catfishes, and feather backs were found to be most sensitive to the electrified gill net, followed by Indian major carps and minor carps. Nile tilapia required higher current intensities to be entangled in the electrified gill net.

Longer fish dominated in the percentage of catch in the electrified gill net, since a long fish intercept a greater potential difference between head and tail than the smaller ones.

Aluminum conduit electrodes were able to create a larger effective zone than copper wire electrodes, and consequently, increased the catch in the electrified gill net with aluminum conduit electrodes.

Justification for the Claim

Comparative study of some of the prior arts related to the invention of a new three-walled electric gill net to attract and catch more fish

Pub. No. PCT/US 1982/001051 – Method and apparatus for electric fish trawling

Newman James under patent number PCT/US 1982/001051 obtained the patent for method and apparatus for electric fish trawling, which is no way comparable to the invention of three-walled electric gill net to attract and catch more fresh water fish in stagnant waters. Electric fish trawling is a moving fishing gear (a conical bag type of net) which is being dragged in the sea to catch sea fish. Normally trawling is the most important active fishing gear, operated in sea at different depths for catching fish. In electric fish trawling invented by Newman James, the catch was more due to attraction of fish towards the positive electrode, which was probably placed in front of the trawl net opening.

The present invention of three-walled electric gill net is a static gear placed in water bodies, where other fishing net cannot be operated due to physical obstructions and weed infestations. Normally in conventional gill net the catch is poor as the fishes frequently avoid hanging gill net in the water, unless it is invisible to the fish. As such, the catch per unit effort of conventional gill net is very poor and it takes a minimum 2-3 hours waiting time for some fishes to be caught accidentally.

The three-walled electric gill net, creating a triangular electric field between the mid-panel and the two outer panels, due to the attachment of positive electrode in central panel and two negative electrodes, connected to two outer panels, 42 cm apart from the central panel, fishes in the vicinity of two outer panels, under the influence of electric field near the negative electrodes, tried to move away from the outer panels, and were attracted by the positive electrode of the central panel at the surface. During this process of fright reflex from outer panels and attraction towards the central panel, fishes were gilled and entangled in all the three panels of the electric gill net within a short period (2 – 60 seconds) and were caught.

Pub. No. CN 107996528 - Intelligent electric field net cage fishing device and method

The device and method invented under patent no. CN 107996528 was for entrapping fish in net cages. Three-walled electric gill net did not use any net cage to entrap fish. But it was a modification of convention gill net/trammel net, where the electric field between three panels was used for frightening fishes around negative electrodes and simultaneously attracting them towards the positive electrode. Fishes, in between these two contradictory stimulus were enmeshed and entangled in all the three net panels.

Pub. No. CN 2116349U - Electric attracting machine

The invention relates to attracting fish from water areas, infested with weeds and stones with the help of long handled electrode.

Three panel electric gill net was entirely different from the aforesaid invention. The electric gill net utilized the fish behavior in an electric field, by the way of scaring fish by the negative electrodes of the outer panels and simultaneously attracting them to the positive electrode of the central panel, when fishes were caught by enmeshing and entangling.

Pub. No. GB 879313A - Improvements in electrical devices for catching fish – Smith Res. and Dev. Company Inc.

It was only an improvement which Smith Research and Development Company Inc. have made with respect to impulse generator, transformer and electrodes. These improvements are no way comparable with the three-walled electric gill net as mentioned above.

Pub. No. US 20060265932 - Electric voltage generating fishing lure- Davis, Bret L.

The invention relates to hook and line fishing, where single fish is caught in a hook, and not the net fishing, where many fish are caught at a time. So this invention is entirely different and cannot be comparable with the three walled electric gill net.

Pub. No. US 5644863A – Fishing gear for electric fishing – Verburg; Abraham Leendert

This was a description of a fishing gear used for electric fishing, which consisted of a beam, pair of electrodes over which a potential difference can be generated, as well as, power source and pulse generator. It also described the connection of power source and pulse generator to the electrodes and housing of pulse generator in water tight chamber which was partly flushed with water. No mention was made about using the fishing gear and how the electric field was generated around the gear and how fish were collected after being influenced by electric field (the method for using the gear for fishing). It appears that probably the gear was used to attract and collect fishes. This electric fishing gear is no way comparable with the invention of three-walled electric gill net due to reasons mentioned in earlier paragraphs.

Comparing the previous publications, mentioned above, the following differences could be identified in the new invention of three-walled electric gill net.

The three-walled electric gill net was an improvement of conventional single-walled gill net and three-walled trammel net by interlacing electrodes in all the three panels. Two negative electrodes were attached at the bottom of the two outer panels; while the positive electrode was attached to the central panel, just below the water surface. The triangular field thus created between two negative electrodes and one positive electrode, scared the fishes near the outer panels. Simultaneously, the positive electrode of the central panel attracted fishes towards the central panel. In this process of scaring and attracting, fishes were gilled and entangled in the electric gill net and caught.

The power source may be either a portable AC generator or service line along with an inverter, which convert AC to DC to activate the device.

With the help of step down transformer and variable resistance, the DC out put was reduced to desired level for frightening and attracting fish without injuries of fish due to electrocution.

In conventional gill nets, the catching of fish is by chance and time consuming. Two to three hours time are required for some fish to be gilled or entangled, especially in standing water without water movements. In the electric gill net, short period of electric stimulation (maximum period of two minutes at each spell) drive the fishes to be gilled and entangled, which can increase the catch per unit effort.

Problem Solved by the Invention

Conventional gill net is used to catch every kind of fish from inland waters and sea. The gear in inland waters are stationary, while in sea, it may be stationary or drifting with currents. The catching efficiency of gill nets depend on the movement of fish in the water and is facilitated in waters with mild current and flow. The movement of water (mild current or flow) is absent in stagnant waters, like, reservoirs, ponds and tanks. While drag net or seine could be operated in small ponds and tanks to catch fish, but in large stagnant waters, like reservoirs with un-cleared stumps and boulders at the bottom, dragging or trawling is not possible. There, gill net is the only alternative to catch fish and the catch in gill net depend on the movement of fish. As such, the catch per unit effort by gill net in stagnant waters remain low. Fish also avoid gill net (a wall type of net) in day light.

In the electric gill net, electric stimulation through conducting electrodes, attached to the net, is provided to fishes between three panels parallely placed gill nets, where a suitable electric field was established between positive and negative electrodes. Fishes around negative electrodes in two outer panels were frightened by electric stimulation, moved forcefully, got attracted to the electric field around the positive electrode of the central panel. Both these repelling and attracting stimulus drive fishes towards the electrified gill net panels and were either gilled or entangled in the net webbings.

Solutions Provided by the Invention

Pulsed DC of low frequency could effectively stimulate fish to repel from the electric field near negative electrode; and attract to the electric field near positive electrode. These attracting and repelling stimulus forced the fish to move away from the negative field, created near the two outer panels of three-walled electric gill net. Being scared from the negative field, fishes were forced to swim towards the positive field around the central panel of the electric gill net, During these forced movement, fishes were either gilled or entangled in the meshes of three panels and caught.

Advantages

1. Higher catch per unit effort in electrified gill net is possible, if the net panels are properly spread and fishes are effectively stimulated.
2. Operation time of electric gill net could be minimum, to the duration of electric exposure.
3. The electric gill net could be shifted to a number of areas for more coverage, which could increase the catch.
4. Fishes caught in the electric gill net would be in prime condition, as there would be no spoilage due to delayed lifting of the net as it happen in case of conventional gill net.

5

Development of a Prototype Electric Dip Net

Principal Investigator – Dr. K.P. Biswas

Associate in the Work – Miss. Shaik Kouser

Background of Invention

Untold centuries ago, some wandering Indians, or his aboriginal predecessor discovered that he could catch salmon at the falls. In the Columbia River, a few miles below the mouth of the Deschutes River, virtually nothing is known about the beginning of the fishery that developed. Shrouded by the ages that have passed, a flourishing fishery evolved.

Fishing locations were a family inheritance and were passed down from father to son. Disputes concerning fishing rights and other fishery rulings were settled by an Indian group composed of three representatives from each reservation contributing fishermen.

The nets used by the Indians were of two general types. Although they both were dip nets, one was held stationary; while the other moved through the water. The motionless one was referred to as a set dip net, while the other was called a movable dip net or simply a dip net.

The set dip net was attached to a more sturdy pole, usually 3 or 4 inches in diameter and varying in length from 10 to 20 feet depending on the fishing location. The steel hoop varies from 2 to 5 feet in diameter. This was fastened to the hoop by small metal rings, but there was no thong for purse action as with movable dip nets. The net itself was much deeper than the movable dip type; it may be as much as 5 feet deep, while the movable dip net had a bag of only about three feet. The frame of the net was supported in the water by wires connected to the shore or to the platform. Fish were usually caught in this type of net when they were falling back from a falls or a strong current. Occasionally a set net fished in an eddy to catch fish which were resting there. A trigger string was attached to the extremity of the bag net. When a fish strikes the net, the vibration was telegraphed by the string to the fisherman holding the other end of the string and he quickly pulled up the net.

The movable dip net was attached to a pole of about two inches diameter, varying a length from 15 to 25 feet, depending on where it was used. It fished only in a fast current where the water was aerated to such a degree that the visibility of the fish was very limited. The steel hoop is an oval about 30 inches long and 18 inches wide. The mesh varies from two and one-half to five inches, stretched mesh according to the desires of the fisherman. It is attached to the hoop by small brass rings and secured by a leather thong which was held in place by slipping it under a taut twine at the base of the pole. The net was dipped into the water and the current aids the fisherman in forcing it down stream through the rolling water. Other factors being equal, the size of the catch was directly proportional to the speed with which the net passed through the water. The more rapidly it moved, the less chance the fish had of escaping.

While most movable dip nets fished blind, that is, the fisherman cannot actually see the fish he was trying to catch, there was an area at Cedilo where "roping" was practiced. During certain water levels, the river flowed over slanting smooth rocks at a depth of about 1 or 2 feet. When fish swam up over these rocks, they could be seen from the platforms constructed over them. Dippers on their platform waited tensely for a fish to appear, their nets held in readiness. When one was seen swimming up, the net was dipped through the water intercepting the fish. Since only a limited area was suitable for this type of fishing, and it was dependent on exact water flows, this method was used to only a minor extent.

With the movable type of dip net, fish moving upstream were usually captured by entering the net head first. When a fish stroke the net, the thong pulled free and the net pursed around the fish preventing its escape. The pole was pulled in hand over hand and lifted to the platform on which the fishermen was standing, which was usually 3 or 4 feet above the surface of the water. A helper often aided the fisherman in pulling up his dip net and in dispatching fish. During the height of the run, two or three fish were sometimes caught in one dip net (Schoning, *et al.*, 1951)

While studying the randomness of rainbow trout samples collected by dip net from tanks (Thorburn, 1991) three methods had been adopted. In one method, samples were dip-netted near the water inflow without prior disturbance of the fish. In the second method, samples were dip-netted from the center of the tank immediately after stirring its contents. The third method involved concentrating fish by releasing water from the tank and then sweeping a dip net through them. Without prior disturbance, larger numbers of the non-crowded population tended to be caught near the water inflow; following mixing smaller fish were caught more frequently.

Foreign in origin and commonly known as Chinese dip net were reportedly introduced in Kerala by Portuguese. This stationary fishing gear, having a mesh size of 12 to 14 mm are fixed in the tidal regions of the inshore waters with sinkers, floats and stakes and are operated up to a depth of 5 meters. The gear is operated from a wooden platform using lever system and is lifted at short intervals by using a 5 h.p. engine. The net is connected to ropes tied with boulders for balancing the net, so that it can be dipped and lifted at short frequencies of 10-15 minutes based on the availability of catch. Net operations depend on tide, as it is made functional during low tide and fishing is stopped as the tidal level rises. At night, a lantern hung from a rod used to attract fish. Once the net is lifted, fish are removed using a scoop net. Maximum species diversity was reported in May and minimum in December. High species diversity found during May-July relates to southwest monsoon period, when fishes migrate to the backwater for their nursery phase. Since the majority of the species spend a part of their life cycle in the backwaters (post-larval and juvenile phase), they are easily caught in the dip net (Lakshmi Pillai, 2016).

From the information available about all the three types of dip net operated in different regions, the catching ability of the dip net largely depend on the movement and migration of fish, initiated by the movement of waters, either as flowing and rolling waters in rivers and streams or on the tidal water movement. The Chinese dip net in Kerala is made functional during low tide and catching of fish is stopped during high tide. The scooping of rainbow trout yearlings from tanks also required movement of water in the form of inflow, stirring and releasing the water from the tank, which acts as a stimulus for the fishes to congregate in a place to be caught by dip net. In Chinese dip net a lantern was reported to be hung over the net at night to attract fish in the netting region.

With these available informations, the idea of a prototype electric dip net was conceived, where fishes will be gathered by attracting them towards the net through suitable electric stimulation.

The Invention

A dip net is a bag-shaped net attached to a rigid frame. The dip net is operated in many coastal areas (estuaries and backwaters). Some dip nets are square, others are round in shape. Frames can be of metal or bamboo. The conventional dip nets used in India are (a) Hand dip net and (b) Chinese dip net. These nets are used in shallow tidal waters to catch small sized fishes, when they gather in an area or over the net.

(a) Hand Dip Net

60 cm X 60 cm, suspended from the ends of two curved sticks, which cross at right angles in the middle. A cord is tied to the crossing of the sticks and the net is lowered vertically into the water. A strong smelling bait is tied at the center of the net to attract crabs and prawns. The mesh size of the net is 1.25 cm and is operated by one or two persons.

(b) Chinese Dip Net

It is a 10 to 11 meter square shallow bag net with a narrow cod end. The huge net is stretched from crossed bamboo poles in the same manner as in the hand operated dip net. For operation of the net, a complicated counterpoised frame work is necessary. The net is vertically lowered into the water and raised. All type of surface swimming fish are caught in the dip net.

The catch per unit effort of the conventional dip net is very poor as small fishes assemble over the net during low tide to nibble on the submerged net, since no lure is kept over the net.

The use of electric current in water to stun fish for capture is a powerful tool for sampling technique for nearly a century. Scientists working with fish and other aquatic organisms are using this technique for fish population studies in rivers and streams, especially, in case of endangered and threatened ones that are impacted by anglers and fishers.

But the technique involves a very dynamic complex of poorly understood mix of physics, physiology and fish behavior.

The effect of electric current on the fish can be categorized under;

1. Galvanotaxis – attracting the fish towards the positive electrode (anode),
2. Frightening – blocking effect of negative electrode (cathode),
3. Narcotizing – hypnotic effect.

The cumulative results of many years of individual piece meal investigations are now known to the responses of fish in electrical field.

In mid-fifties, many researchers put emphasis on current density of the electric field to elicit different responses of fish, which together with voltage gradient between head and tail, referred to as "body-voltage" was thought to be the cause of involuntary taxis of the fish. Orientation of the fish's body axis with relation to lines of current conduction was reported to initiate perception and movement.

In a homogeneous electric field in water, the current density at all its points is the same. The field pattern consists of equally spaced parallel equipotential planes. As the current spreads into the water around any electrode, the current density drops off steadily as the distance from the electrode increases.

Every electrode therefore, has a near zone of high current density, which may be above the danger level. Surrounding the nearest zone, is a region in which the current density is in the effective range. The current density gradually drops off until it is no longer large enough to cause the desirable reactions and this makes the end of what might be called the "effective zone". Beyond this distance, the current density continues to drop until it is no longer above the perception level. An effective electrode is one, which has a very small danger zone, a maximum effective zone and the smallest possible perception zone.

The study was undertaken to determine the possibility of increasing the catching capacity of a dip net by utilizing the attracting effect of the anode and frightening effect of the cathode to assemble fish in a hypnotic state over the dip net and thereby increasing not only, the catch per unit effort, but also to catch large sized fishes, which the conventional dip net failed to do.

The prototype electrical dip net was a conical net with wide circular mouth (0.85 m) and a narrow cod end (0.35 m). The net was made of multifilament nylon twine No. 4 having an uniform 2.8 cm stretched mesh (knot to knot).

A stainless steel hollow sphere of 9 cm diameter was used as positive electrode, which was kept immersed under the water surface. The hollow globe was filled with measured quantity of water to obtain neutral buoyancy, so as to keep the globe immersed above the center of the wide mouth of the dip net, just below the surface. Two one meter 14 gauge un-insulated copper wire in the shape of hemispheres were attached along the diameter of the wide mouth (0.85 m) and was electrically connected to each other to serve as cathode.

In the second set up, only positive electrode was changed with an aluminum ring with 8 droppers and the same 14 gauge copper wire hemispheres were used as cathode. Aluminum ring of 5 mm thickness having a diameter of 14.3 cm was electrically connected with equally spaced 8 vertical aluminum rods of 20 cm length (droppers).

In the third set up, the positive electrode consisted of one 20 cm diameter coil of aluminum conduit. The outer diameter of the conduit was 1.2 cm, which was kept immersed under the water surface over the center of the dip net mouth. Two one meter bent aluminum conduit in the shape of hemisphere were attached along the wide mouth of the dip net, electrically connected, served as negative electrode.

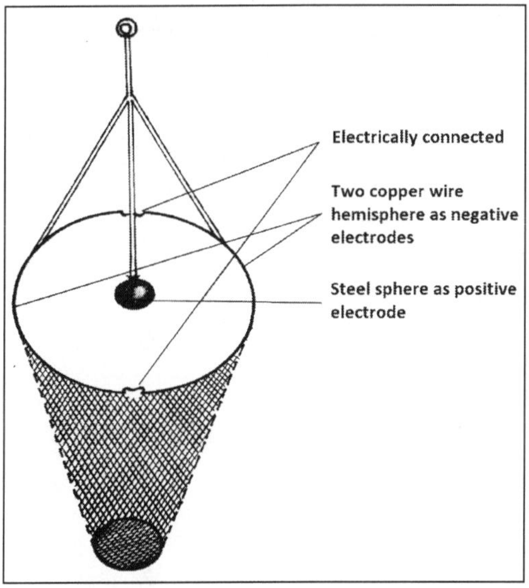

Schematic Diagram of Electrified Dip Net Showing the Position of the +Ve Electrode (Steel sphere).

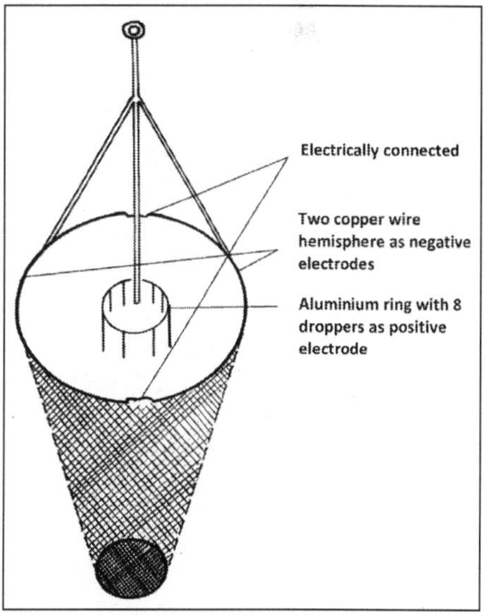

Schematic Diagram of Electrified Dip Net Showing the Position of the +Ve Electrode (Aluminium ring).

A tangential vertical field in the shape of a cone was created between anode and cathode, when the net was energized electrically.

The specification of the conventional dip net, used for comparison with electric dip net was similar to that of electric dip net, except the electrodes, that is, a conical shaped net with a wide circular mouth (0.85 m) and a narrow cod end (0.35 m). The dip net was made of similar netting materials and of same mesh size as of electric dip net.

With the help of cords attached to the net mouth and a stick, the net was lowered and lifted vertically from the water. The operation and hauling of the conventional dip net was maintained similar to that of electric dip net.

Experiments were conducted in a glass tank of 1.8 x 1.05 meter, filled with water to a depth of 0.52 meter, having a temperature range of 22 to 31 degree Celsius and electrical resistance of 5000 to 6700 ohm/sq.cm.

A 0.5 kVA, AC to DC inverter, capable of supplying 40 to 230 volt direct current was used to energize the dip net. An electronic pulser of low frequency (1/sec to 3/sec) was connected between the inverter input and the electrodes to obtain pulsed direct current (PDC).

Nine species, of fish, commonly available in Indian freshwater were used in the trials of electric dip net. They were *Labeo rohita, Cirrhinus mrigala, Labeo calbasu, Labeo bata, Notopterus notopterus, Cyprinus carpio, Oreochromis niloticus, Channa punctatus, Heteropneustis fossilis.*

In the trials, round conical net with wide circular mouth of 0.85 m diameter was used. The narrow cod end (0.35 m diameter) framed on a galvanized wire of 3 mm thickness was closed at the time of operation.

A known number of multiple fish species, which are very common in Indian freshwater tanks, ponds and reservoirs were utilized to ascertain their reactions towards the electric dip net and conventional net without electric connection.

The same dip net without electrical connection was used as conventional dip net to have a comparison between the electrified dip net and conventional type with respect to their catching efficiency, that is, catch per unit effort (CPUE) and the time taken to immobilize and catch fish.

Water temperature and electrical resistance of water of the experimental tank were measured and recorded at the beginning of each set of trial.

The conventional dip net without electric connections was lowered vertically at the center of the tank bottom, so that the wide mouth of the reinforced 0.85 m galvanized iron wire ring settle at the bottom. The net was allowed to remain in the bottom of the tank for 30 minutes for the fish to acclimatize with the presence of the net in the water. After 30 minutes the net was lifted with the help of a bamboo stick to which the rim of the net was tied loosely. The net was lifted suddenly above the water surface, so that fish over the net could be caught on surprise. The catch was then counted, measured for length, species-wise and the fishes were released back into the tank to maintain the fish population as before.

The electrodes (positive and negative), fitted in the net, but not electrically connected earlier, was then connected to the DC power source of a regulated voltage in such a way, so that the negative terminal of the power was connected to the electrodes around the wide rim of the net mouth, which settled at the tank bottom. The positive terminal of the power was connected to the floating metal sphere, which was positioned centrally over the negative electrode under the water surface. Energizing the electrodes with DC power source, a conical tangential vertical electric field was established, the intensity of which depended on the potential difference applied between the electrodes.

Since dissimilar electrodes (positive and negative electrodes) were used in the trial, the electric field that was established between the anode and cathode was heterogeneous and was necessary to determine the electric field pattern between the positive and negative electrodes at a given voltage.

For measuring the current profile between positive and negative electrodes, a silver probe of one square centimeter was used to measure the density of current and voltage gradient that was established at a given potential difference.

Three different sets of electrodes (galvanized iron sphere as positive and bare copper wire as negative; aluminum ring with droppers as positive and bare copper wire as negative; and aluminum conduit coil as positive and aluminum conduit hemisphere as negative) were used in the trials to compare their efficiency over others.

After setting the net at the tank bottom, electric current was allowed to flow through the electrodes fitted to the wider mouth of net rim and suspended in the water surface over the net centrally. Fishes within the effective range were found to rush over to the positive electrode and immobilized. The flow of current to the net was discontinued as soon as the suspended anode at the surface was out of water. Fishes collected in the net was counted, measured for length, species-wise and released back in the tank with heavily oxygenating the water with a powerful air pump.

When fishes were fully recovered from the stress within two hours, the trial was repeated with different electrical strength and with different sets of electrodes. Three trials in a day was made with varying field intensity (starting from lower to higher order) and the fishes were given a rest period of 48 hours before the next trial.

Fishes of different species combination, size group were used in the trial in addition with the fishes of same species individually, duo and trio for studying their response in depth.

All electrical parameters during the trials were measured with two precision multi-meters (AVO, Universal, Model 8 and Sanwa-P-2).

Significant Findings

As the electrodes (positive and negative) were fitted in such a way that the positive electrode will always remain immersed under the water surface over the central point of the dip net and the negative electrode along with the rim (mouth) of the net at the bottom, a tangential vertical electric field was formed when the net's electrodes were connected to the output terminals of power supply during trials of the electrified dip net. The field that was created was heterogeneous and was difficult to measure because of the inclination of lines of current conduction in a parabolic path, due to different shapes and size of anode and cathode fitted in the net.

In all the three sets of electrodes, used during the trials (steel sphere and copper wire hemisphere; aluminium ring with dropper and copper wire hemisphere; aluminium conduit coil and aluminum conduit hemisphere), the voltage gradient (V/cm) was found to decrease with the increase of the vertical distance from the positive electrode. The values of the voltage gradient was 33.1 V at distance of 4 cm away from the positive electrode, which progressively decreased to 24 V at 8.5 cm, 16.1 V at 18.5 cm, 11.5 V at 28.5 cm, 10 V at 38.5 cm and 9 V at 39 cm away from the positive electrode, when a galvanized iron sphere and copper wire hemisphere were used as anode and cathode respectively.

The same patterns of voltage gradient were observed with other two sets of electrode combinations. With aluminium ring and droppers as anode and copper wire hemisphere as cathode, the voltage gradients at 4, 12, 20, 30, 40 and 41cm away from positive electrode were, 25.0, 19.5, 16.0, 12.1, 11.0 and 10.0 (V/cm) respectively. In electrodes of aluminium conduit coil and aluminium conduit hemisphere, however, voltage gradients of 32.0, 20.0, 15.0, 10.0, 9.0 and 8.0 V/cm were recorded at distances of 4.0, 12.0, 20.0, 30.0, 40.0 and 41.0 cm respectively from the anode. Though the voltage gradient in near positive zone (4 cm away from positive electrode) created by aluminium ring with droppers (anode) and copper wire hemisphere (cathode) was less (25.0 V/cm) than the voltage gradient established by steel sphere (anode), copper wire hemisphere (cathode) and conduit coil (anode) and conduit hemisphere (cathode) 33.1 and 32.0 V/cm respectively, yet the range of the electric field from ring with droppers was wider (4 - 41cm), than the other two combinations where the voltage gradient (10 V/cm) were limited within 38.5 and 30 cm away from positive electrode in sphere and copper wire hemisphere and conduit coil and conduit hemisphere combination respectively.

The voltage gradient, when measured for horizontal field, with these three electrode combinations for a distance of 10 to 60 cm away from the positive electrode, the aluminium ring with droppers combination exhibited slightly lower voltage gradient (25 - 9 V/cm) than the sphere and conduit coils by 33.1 – 9.0 V/cm and 32.0 – 8.0 V/cm respectively. The voltage gradients in all the three electrode systems decreased (9.0, 9.0 and 8.0 V/cm) as the horizontal distances from positive electrode increased 10 - 60 cm.

The strength of current density (mA/cm^2) between positive and negative electrode was checked by increasing the voltage at the electrodes, which ranged 0.02 mA/cm^2 at 6 volts to 8.5 mA/cm^2 at 220 volts. The current density progressively increased with the rise of input voltage of 9, 11, 15, 24, 32, 43, 58, 80, 120, 130, 135, 155 and 220 volts to 0.1, 0.15, 0.25, 0.45, 0.75, 0.85, 1.0, 1.6, 2.3, 4.9, 5.8, 7.0 and 8.5 mA/cm^2 respectively. It followed the same rule of electric current conduction through the water (electrolyte). This was necessary to determine the safe limits of current intensity to be applied during trials without causing as little harm as possible to the fishes.

The percentage of fish catch in electrified net was determined by application of different voltage range (18 – 200 V), consequently with the rise of current density from 1.6 to 8.0 mA/cm^2. The time of electrical exposure was restricted between 19 – 90 seconds. At each voltage and current density strength, the number of fish caught in the electrified net was recorded and the percentage of capture out of total population was calculated.

It was observed that the non-electrified net was able to catch 4 - 13 per cent of total fish population and in one trial even nil catch in others, where as the electrified dip net was able to attract and catch 27 – 100 per cent of total fish population at different combinations of electrical parameter and time of exposure to the fish. It was interesting to note that the percentage of fish catch in electrified net increased with the increase of voltage and current density up to the level of 190 V and 7.5 mA/cm^2, when the total fish population (100 per cent) could be attracted and captured in the electrified net. With the further rise of potential difference to 200 V and current density to 8 mA/cm^2, the catch per unit effort dropped down to 79 per cent. This indicated that for the electrified dip net, 190 V and current density of 7.5 mA/cm^2 was optimum, and no appreciable significance had been observed in exposure time. The attraction of fish to the anode was best at a current density of 7.5 mA/cm^2 and the higher current density beyond this have caused seizure of fish muscle, particularly those, swimming nearer to the effective zone.

Fish Catch in Electrified Dip Net.

The picture was different, when the sphere anode had been replaced by an aluminium ring with eight droppers of aluminium rods. In this case of current density of 4 mA/cm^2 was found to be most effective (90 per cent capture), when the applied voltage was 175 V and the exposure of current was limited to 30 seconds. In higher voltage of 200 V, when the density of current increased to 8.5 mA/cm^2, 85 per cent of fish population were attracted and caught by electric dip net.

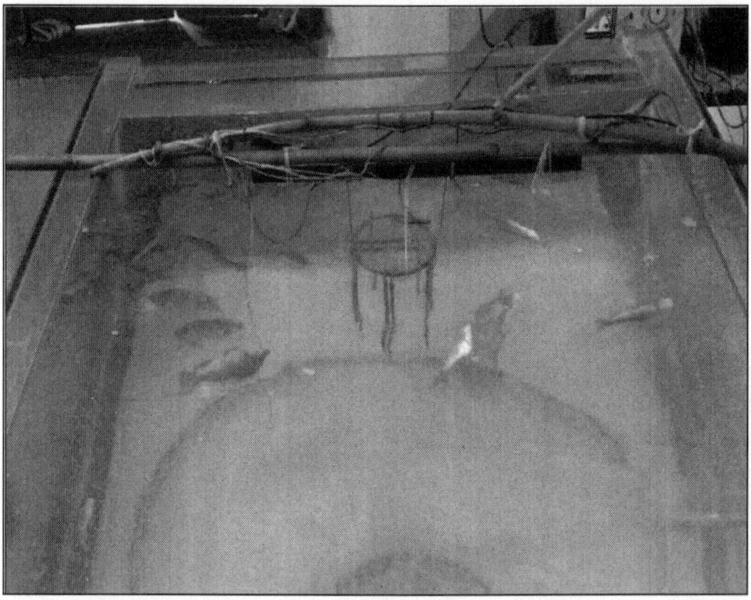

Gathering of Fish over Electrified Dip Net.

The voltage below 175 (120 – 145 V) could catch 37 – 88.4 per cent of fish when the exposure time was 17 – 30 seconds. Therefore, in this electrode set up (anode with aluminium ring with 8 droppers) 4.0 to 6.4 mA/cm^2 (current density) was considered to be optimum for maximum catching efficiency of electrified dip net.

Changing both anode and cathode with aluminium conduit and shaping the anode as one and half coil of 20 cm diameter and two 40 cm bent aluminium conduit to form a hemisphere and connected together as cathode, the electrical field pattern completely changed. By applying a voltage of 55 to 145 V only, a current density of 0.075 to 0.15 mA/cm^2 could be established. As a result, the percentage of total fish population that could be attracted and caught in electrified dip net was very less (7 to 33 per cent) even though the electrical exposure was applied for 19 to 58 seconds. The non-electrified net however, was unable to catch any fish.

When the dip net was stationary in one place, lying at the bottom of the tank, the number of attracting fish to the net, when electrified, was limited to a maximum of 37 per cent in case of anode with ring and droppers and 2.5 to 27 per cent when the anode was aluminium conduit coil. But when the net was moved horizontally at a velocity of 120 cm/min the percentage of fish catch increased substantially (53–90 per cent) in ring with droppers anode and 36–87 per cent in aluminium conduit coil positive electrode.

Fishes were found to be frightened and took refuge in no influence zone during electrical exposure when the net remained in one place (stationary). The escaped fishes were influenced and captured when the net along with electric field moved slowly in water, horizontally, indicating the higher catching efficiency of mobile electric dip net.

The continuous DC was found to be effective to catch 34 – 37 per cent in stationary electric dip net and 39 – 73 per cent in mobile electric dip net. PDC (pulsed direct current) 1/sec was found to be better than PDC 2/sec and PDC 3/sec. In PDC 1/sec, 20 per cent fish was caught in stationary net and 100 per cent in mobile net. The next effective pulse rate was 3/sec, when it could catch 7 per cent of fish population in stationary and 47 per cent in moving net by applying 70 – 140 V at current density of 0.09 to 0.22 mA/cm^2.

Specific reactions of individual species like *L. rohita, C. mrigala, L. bata, O. niloticus, H. fossilis* and *C. puntatus* towards electric dip net was ascertained with three electrode systems. 100 per cent of *L. rohit*a and *L. bata* were caught in electric dip net at 110 – 145 V generating current density of 0.02 – 0.15 mA/cm^2. Higher voltage of 165 – 170 V producing 0.16 - 0.165 mA/cm^2 was required to capture 100 per cent *O. niloticus*. 72 – 84 per cent of *C. mrigala* was caught when the voltage was 110 – 140 V producing 0.1 – 0.15 mA/cm^2. In all these cases 2 – 10 seconds of electrical exposure were applied.

In steel sphere and copper wire electrodes, at 117 V producing a current density of 0.125 mA/cm^2, 100 per cent *C. mrigala, H. fossilis* and *C. punctatus* were caught in 2 seconds. 89 per cent of *L. rohita* and 75 per cent of *O. niloticus* were caught within 3 – 8 seconds in this electrodes set up.

With ring and droppers and copper wire electrodes, among three species, namely, *L. rohita, O. niloticus* and *C. punctatus* between 124 – 185 mm total length, *L. rohita* was influenced by 0.095 – 0.45 mA/cm^2 and *C. punctatus* by 0.05 – 0.095 mA/cm^2, while *O. niloticus* required 0.1 – 0.15 mA/cm^2 to be immobilized and caught. *H. fossilis, C. puctatus* was most sensitive to electric field for immobilization, followed by carps, namely, *C. mrigala, L. rohita* and *L. bata. O. niloticus* was less sensitive to electric stimulation and required higher current strength for immobilization and capture.

In different current densities, the effective area from positive electrode was shifted. In lower intensity (0.04 mA/cm^2) the range of effective area was 46 cm from the positive electrode, while in higher intensity (0.05 mA/cm^2) the distance was increased to 70 cm from the anode. 100 per cent anodic attraction and capture could be effected at 0.04 mA/cm^2 when they were 46 cm or less than 46 cm away from the positive electrode. Beyond 46 cm from the positive electrode no fishes were influenced and 100 per cent escaped from the field. In slightly higher current strength of 0.05 mA/cm^2, the distance from the positive electrode increased to 52 cm for attraction and immobilization of 100 per cent fishes. 60 – 76 per cent of fishes were influenced when they were between 57 – 70 cm away from the anode. No fishes were found to be effected when they were beyond 70 - 80 cm away from the positive electrode. This variation in effective zone with rise of field intensity had an important bearing in attracting and immobilizing fishes in electric dip net.

A group of four *L. rohita* were attracted and caught in electric dip net in the smallest range of current density (0.004 – 0.005mA/cm^2), *C. mrigala* in a group of those fishes, exhibited cent percent attraction and capture in the electrified dip net at current densities of 0.01 – 0.015 mA/cm^2 onwards. 50 – 75 per cent *L. bata* escaped at a lower density of current ranging from 0.006 – 0.3 mA/cm^2. Only after increasing current density to 0.446 – 0.45 mA/cm^2 by applying higher voltages fishes responded for 100 per cent electrotaxis and capture. All *C. punctatus* of 180 – 185 mm were subjected to forced swimming towards electrified dip net and were caught in narcotized condition at current density of 0.001 -0.006 mA/cm^2. A little higher current density (0.021 – 0.025mA/cm^2) was required to attract and immobilize all *H. fossilis* to be caught in electrified dip net.

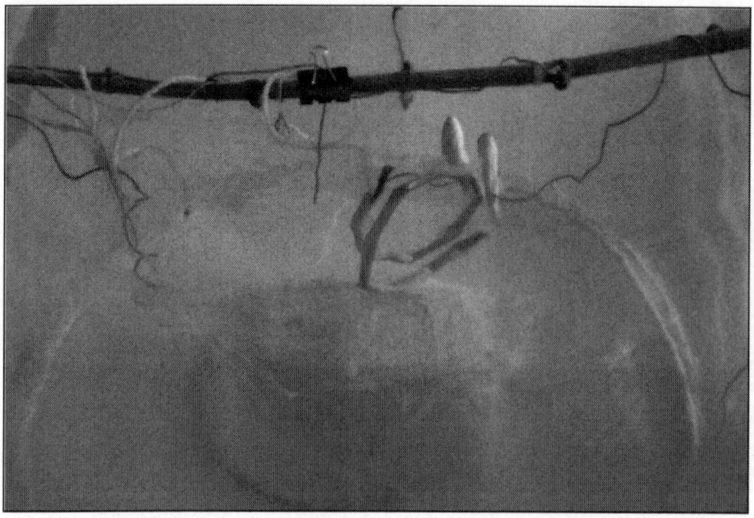

Reaction of Fishes towards the Electrified Dip Net.

Forty percent of *O. niloticus* escaped out of the field and the electric net at a current density of 0.056 – 0.075 mA/cm^2 and 100 per cent attraction to the net and capture was possible when the field intensity was raised to 0.121 – 0.125 mA/cm^2.

Most of the fish species responded from current density of 0.035 – 0.04 mA/cm^2 developed by the application of 51 – 70 volts. With the rise of density of current (0.04 – 0.075 mA/cm^2) generated by 71 – 170 volts, the increases in percentage of attraction and capture in electric dip net was noticed. With slight variation among species and the field strength, 100 per cent capture of four species, namely, *L. rohita, C. punctatus, H. fossilis* and *O. niloticus* had been recorded when the density of current was 0.08 mA/cm^2 produced by 190 volt.

In a field of 0.06 – 0.16 mA/cm^2 generated by the application of 95 – 200 volt DC current, 100 per cent of *L. rohita, H. fossilis, O. niloticus, C. mrigala* were caught in electrified dip net when they were between 151 – 170 mm TL. Highest percentage (84 per cent) of *L. bata* of 151 -170 mm was attracted to the electric net and caught.

Fishes of all the six species did not swim to the electrified net and caught unless their total length were more than 150 mm.

Claims for the Invention

The dip net is a passive fishing gear, used in small scale fisheries and is known as chance fishing of wait and see. The conventional dip net used in India include a hand dip net small size (60 X 60 cm) operated by one of two persons and Chinese dip net of 10 – 11 meter square, operated from a cross bamboo frame to lower and lift the net in the water (Biswas, 1996).

The prototype electric dip net was designed to impart electrical stimulus to the fishes, sufficient enough to assemble involuntarily over the net, immobilized and caught when the net will be lifted out of water.

The electric field generated vertically and tangentially through the water columns by the application of DC or PDC will induce the fish to swim towards the anode forcefully. As they approached the positive electrode, get stunned on reaching the effective zone over the dip net, unable to move and caught in the net when lifted.

As the electric current supplied to the electrodes passed into the water in all directions created an electric field. The current and voltage associated with the electrode is distributed through the water and the nature of this distribution is critical to effective electric fishing (Cowx and Lamarque, 1990). The electrical parameters used to describe the distribution of current were the current density (mA/cm^2) and the voltage gradient (V/cm). The quantities are measures of the strength of the current and potential difference at a point in the distribution and are related to each other by the electrical conductivity or electrical resistance of the water.

As the current spreads into the water around any electrode, the current density correspondingly dropped off steadily, as the distance from the electrode increased. Every electrode, therefore has a near zone of high current density which may be above danger level. Surrounding the nearest zone is a region in which current density is in the effective range. The current density further dropped off with the increase of distance from the electrode until it is no longer large enough to cause desirable reactions (end of effective zone).

The distribution of current and voltage in the water near an electrode depend on the size, shape and spacing of electrodes, besides the material with which the electrodes are made of.

Isolated spherical electrode emits current radially, cutting by the concentric constant voltage lines. The spherical electrode produces lower current densities in the danger zone, higher current densities in the effective zone (mid-region) and lower current densities in the far region (perception zone). These effects are results of the more uniform spreading of current around spherical electrode and lead to the generally superior performance of spherical electrodes.

In case of cylindrical electrodes, the current density is uniform along the length of the cylinder and hence falls off more slowly in proportion to the first power of distance. Due to these differences, the voltage distribution and hence the resistance of the two shapes are quite different.

Various types of electro-fishing equipments were used throughout the world for various purpose, mostly for fish population census and diversion of fish population for breeding in lentic and lotic waters.

In all these electro-fishing investigations fishes were influenced to swim around the shocker or catcher, stunned and collected by the netting party who followed the electro fisher. In this method, some fish in the border of effective zone escape, hide under cover.

In the prototype electric dip net, the electrode floating over the center of wide mouth of the net, forced the fish to assemble in the middle of the net due to the influence of a triangular electric field created over the net mouth up to the surface.

Irrespective of three sets of electrodes used in the trials, fishes of different species were found to swim towards the dip net and caught. More the density of current prevailing in the field, higher the percentage of fish population caught in the net. When non – electrified net could catch only insignificant numbers, ranging from 3 – 13 per cent of the total population, 37 – 90 per cent fishes were caught in electric dip net, when aluminium ring with droppers was used as anode.

Fish assemblage in Chinese dip net depend mainly on the tidal movement of water. In addition artificial lights and baits are used to attract fish over the net. Fisher has to wait for sufficient fish gathering over the net before lifting the net for catch. Electrified dip net in stationary condition could catch 37 per cent and 27 per cent with ring and droppers and aluminium conduit coil as anode respectively. The same net registered a higher catch of 53 – 90 per cent and 39 -87 per cent respectively when the net was moved horizontally at the rate of 120 – 180 cm/min.

In stationary condition, DC was found to be superior since, it collected 34 – 37 per cent of fish in the net. In PDC 1/sec only 20 per cent fish was captured. In the horizontally moving net, however, PDC 1/sec could collect 100 per cent fishes, followed by PDC 3/sec (47 per cent). The performance of DC was improved in catching 39 -73 per cent depending on the strength of current density (0.09 – 0.22 mA/cm^2) when the dip net was drawn horizontally.

The electric dip net could attract and caught 100 per cent of fish population from a distance of 46 cm downwards from the positive electrode in lower current density of 0.04 mA/cm^2. Increasing the current strength to 0.05 mA/cm^2 the distance of anodic attraction could be increased to 70 cm. For cent percent fish capture without any fish escape, 57 – 70 cm from the positive electrode was found to be optimum.

As regards the specific reaction 100 per cent attraction and capture in electrified dip net was effected in *L. rohita, C. punctatus* and *H. fossilis*. But species like *O. niloticus, C. mrigala* and *L. bata* were attracted and caught in electric dip net between 50 per cent and 75 per cent. Specific sensitivity of fishes towards electric field do exist, which was also indicated by Meyer-waarden, 1957, Biswas, 1971a.

Increasing current density from 0.0045 – 0.05 mA/cm^2 influenced fishes of all the species taken up for trials, to a varying degree. In lower intensities of 0.02 – 0.03 mA/cm^2 only three species, namely, *L. rohita, N. notopterus* and *L. calbasu* responded positively. Possibility of selective fishing with electric dip is promising by regulating field strength.

Halsband (1954) and Meyer – waarden (1957) reported that the values found for narcotizing threshold are not to be considered absolute, but differ considerably for the various species as well as for fish of varying sizes. The larger the fish, greater is the potential difference between head and tail at certain current density. As such, larger fishes were affected more easily in a homogenous electric field (Halsband, 1954; Meyer-Waarden, 1957, Denzer, 1956, Biswas, 1971, a, b, c, 1974, 1976). The electric dip net produced heterogeneous electric field in all the three electrode combinations, where the electric field between the electrodes were not uniform. 100 per cent attraction and capture in electric dip net was observed in case of *L. rohita* of 151 – 190 mm total length. All *C. punctatus* of 191 – 200 mm length were attracted and caught in electric dip. Similarly *H. fossilis, O. niloticus* and *C. mrigala* were caught in electrified dip net when their length were between 151 – 170 mm. Below 150 mm total length, none of the above species had shown cent percent attraction and capture to the electric dip net.

Therefore, the prototype electric dip net clearly indicated its superiority over the conventional dip net due to following reasons;

- ☆ Higher percentage of fish catch in electrified dip net due to larger effective zone created by aluminium ring with droppers as anode.
- ☆ As regards type of current, DC field was found superior, followed by PDC 1/sec and PDC 3/sec.
- ☆ An effective zone of 46 cm from positive electrode was found to be most effective range with this prototype dip net. The effective range can be suitably increased by increasing the dimension of the net, size and arrangements of electrodes.
- ☆ Among nine species of common Indian fresh water fishes, *L. rohita* followed by *N. notopterus, C. punctatus, H. fossilis, C. mrigala* were found to be most susceptible towards the electric dip net. *O. nilotica* needed higher current intensity for capture.
- ☆ Fishes above 150 mm length, irrespective of varieties, can be influenced towards electric dip net.
- ☆ Slowly moving electrified dip net gave a higher fish catch than that of the net located at one place.

Justification for the Claim

Comparative study of some of the prior arts with that of development of a prototype electric dip net

Pub. No. CN207125179U - Quick fishing net device of electric work

The invention on the patent no CN207125179U concerning quick fishing net device of electric work is a luminous function, to attract fish to assemble near the source of light when they are caught.

The prototype electric dip net for which the patent has been claimed is a different device where fishes are attracted by the positive electrode when a suitable electric field was created between the centrally suspended anode and outer rim of cathode. As such the present invention is quite different from that of the device of aforementioned patent.

Pub. No. GB2001207 - Apparatus for the transmission of electrical signals on a fishing net

The apparatus for the transmission of electrical signals on a fishing net relates to the position and movement of the fishing net under water, mainly for knowing the proper functioning of the net under water.

The prototype electric dip net is entirely different from that of the aforementioned patent.

Pub. No. US2012019907AI - Electromagnetic apparatus and methods for affecting behavior in elasmobrach fish

The electromagnetic apparatus and methods for affecting behavior in elasmobranch fish under the patent no US2012019907AI is a part of a large system and claimed to be incorporated in the variety of electromagnetic waves, required to study the behavior of elasmobranch fishes like rays, skates and sharks in the sea. The device is not at all comparable to the prototype electric dip net for which patent has been claimed.

Pub. No. US2004017426AI - Electrically monitored fish farm net and method

Electrically monitored fish farm net and method under the patent no US2004017426AI relates to monitor escape of fish from fish net cages where fishes are cultured in restricted way in the water. This invention is no way comparable to the prototype electric dip net where it has been designed to catch more fish in the dip net.

Pub. No. US20050091905AI - Lighted fish landing net

Lighted fish landing net under the patent no US20050091905AI relates to fish friendly environment for avoiding scaring of fish, used with a fish landing net, mainly because to avoid spilling of fish from the net. This is no way comparable to the prototype electric dip net.

Pub. No. 2006016887AI - Electrical control device for marine animals

Electrical control device for marine animals under patent no 2006016887AI is a device, used for temporary paralyzing marine animals for safe handling by human for bringing it to water craft without damaging the animal.

This is entirely different from that of the prototype electric dip net.

Pub. No. US10051850BI - Apparatus and methods for electro-fishing

Apparatus and methods for electro-fishing under the patent no US10051850BI is an equipment for generating power transfer for a targeted fish species which can be used in electric fishing system. It is only an equipment required for the study. So, this cannot be comparable to the prototype electric dip net.

Pub. No. CN1045866C - Middle and high voltage electric fish barrier net with two rows and three electrodes

Middle and high voltage electric fish barrier net with two rows and three electrodes is an electric barrier to prevent migration of fish and is applicable to the spillway of rivers, lakes and reservoirs. The device is checking the movement of fish from a particular area. It cannot be compared with electric dip net which was designed to catch more fish.

Pub. No. CN2355535Y - Electric net for tucking fish

Electric net for tucking fish under the patent no. CN2355535Y describe the different component such as power transmission line, supporting frame, floater and pulse power source which claim to be suitable for occasions where fish is driven in a fish pond, a water storage reservoir and river. The device is no way comparable to the electric dip net for which patent has been claimed.

Comparing with the previous publications mentioned on the search report, the following differences have been identified in the present invention "Development of a prototype electric dip net" for which patent has been applied:

1. The electric dip net is an improvement of the indigenous dip net (Chinese dip net) where the dip net is kept immersed in the water and one has to wait for considerable time to assemble fish over the dip net. Normally small fishes come over the dip net for food and hide out. After a considerable time, the net is lifted and a very small amount of fish are caught by this process. The method is waiting for the fish to be gathered over the net and caught which is very uncertain. Besides, big fishes are seldom caught by the dip net since during lifting the net out of water, large fishes are scared and escape from the net. As such the catch per unit effort is very low.

   The electric dip net utilized the electric field created over at the dip net by the negative electrode at the outer rim of dip net and centrally suspended positive electrode over the dip net causing the fish attracted towards the positive electrodes in a hypnotic condition. Fishes could not escape from the coverage of the net and were caught when the net is lifted.

2. The electric field over the dip net can be regulated by the use of inverter and variable resistance as such the injury of fish due to electrocution is minimum.

3. As soon as the net is lifted from the water the suspended positive electrode came out of water first disconnecting the current flow which caused the disappearance of electric field and the fishes which were already influenced under the electric field were restored under normal condition.

4. Waiting for the fish, assembled over the dip net is minimized due to the attracting effect of positive electrode on fish. As such the net can be operated for a number of times in different places within a short period.

5. Since the large size fish are easily influenced by the electric field the catch will have more number of large size fishes in the electric dip net.

Problems Solved by the Invention

The main problem of the conventional dip net is that it is small in size, used by one person either for subsistence fishing or for sampling from a stock of fish population. In case of sampling, fishes are difficult to capture in required numbers representing the population. In subsistence fishing, fisher has to wait for gathering fishes near about the dip net either during tidal water movement or to use lure (light in darkness) for bringing the fish around the dip net. As such the catch is only a small quantity of small sized fishes as the bigger one will smartly escape from the coverage of the net at the time of lifting. Dip net fishery in the standing water is a chance fishing with a meager catch of small sized fishes like minnows and young ones. The coverage of the dip net is small and the escape in large.

The electric dip net which has been developed, on the other hand, will impart suitable electric stimulus through the water to attract fish towards the net. During this process on crossing the periphery of the net they will be immobilized at the center of the net, unable to move from there, out of their own accords and could be easily caught when the net is lifted out of water. When the net will be lifted out of water, the current flow within the net will cease and fishes could be again salvaged to living condition.

Solution Provided by the Invention

The invention facilitated the capture of more fish by attraction towards the net and reduction of escape out of net coverage, as on reaching over the net they will be immobilized and will remain as such till the net is lifted.

Advantages

1. Increase catch per unit effort.
2. Common Indian fresh water fishes could be easily influenced by the electric dip net and caught.
3. Large sized fishes could be easily attracted to the net and caught leaving the small sized ones.
4. Moving the electric net horizontally, catch can be 2 to 3 folds than of stationary net.

6

Safety Measures to be Adopted in Any Form of Fishing with Electricity

Five inventions described in the preceding chapters of the book may prompt enthusiast researchers to explore the possibility of similar type of work. Fro them, there is a caution, that they should not take up such ventures without required qualifications and training for this type of work with electricity.

The author, Dr. K.P. Biswas started working on electrofisheries only after obtaining special training on electrofisheries under Dr. H. W. Denzer, an internationally reputed electrophysiologist in the State Institute of Fisheries, Albaum, Saurland, West Germany.

It is, therefore, thought to be proper to give a brief idea about the hazards of electrofisheries work and the safety measures, one is required to take before starting this type of work.

Hazards of Electro-fishing: Precautions and Safety Measures

Electro-fishing (passing electricity through water to catch fish) requires sophisticated equipment. While this method is commonly employed by scientists for survey and fisheries management, a crude and illegal form of electro-fishing is employed in some parts of the country, which can result in accidental electrocution and even fatalities.

Safety should be a primary consideration in all electric fishing operations. All personnel involved in electro-fishing operations should be instructed as to the fundamentals of electricity, and understand and observe the safety requirements associated with electro-fishing.

The single most important factor in both electro-fishing efficiency and safety is the training and experience of the crew. Regardless of the safety precautions given, the capability of the crew in adhering to those guidelines and good common sense in handling unforeseen circumstances, is of cardinal importance (Smith,1989).

It is recommended that crew leaders attend a course on " Principles and Techniques of Electro-fishing" either in US Fish and Wildlife Services Fisheries Academy or in State Institute of Fisheries, North Rhine, Albaum, Saurland, Germany.

Hazards of Electro-fishing

Electro-fishing equipment uses voltages and currents that can be lethal to humans. Direct effects of electric shock may include heart failure, respiratory interference or electric burns. There may also be indirect injuries through a worker recoiling violently and striking an object.

The main sources of potential risk of electric shock during electro-fishing operations are:

1. Bodily contact with energized electrodes of opposite polarity;
2. Bodily contact with water within the electric field; and
3. Shocks from inadequately constructed or insulated equipment.

Other hazards include drowning, fire, exhaust gases, and the increased risk of accident due to fatigue.

It is easy for experienced operators to become complacent about the dangers involved in electro-fishing, but they are always present and must be constantly guarded against by regular checks and maintenance to minimize the risk of accidents.

Training and Certification of Operators and Crews

All electro-fishing operations must be carried out under the supervision and control of a Senior Operator who has been awarded a Certificate of Competency in Electro-fishing Procedures and Safety for the particular type of equipment being used, that is, backpack, shore-based or boat-mounted.

Initially, certificates will be awarded to persons who can provide evidence of having experience using a particular electro-fishing gear type for at least 20 sessions (an electro-fishing "session" refers to a single complete electro-fishing operation consisting of gear set up, safety checks, sampling, and gear dismantling.) In this way, several sessions may occur in one day if more than one site is sampled.

The evidence of experience may consist of a signed statement from a University Department, research section, or commercial license authority, plus a current Senior First-Aid certificate or equivalent including CPR, and a recent medical certificate stating freedom from major heart or respiratory complaints that the examining doctor considers could pose a significant risk should even a minor electrical shock occur. Certificate holders will be provided with a manual outlining electro-fishing theory, procedures and safety which they will be required to follow.

The responsibility will then be on these Senior Operators to train and supervise other staff and students, under their control, complying strictly with the requirements outlined in the above manual. All electro-fishing team trainees must have also completed a first aid certificate and pass a medical examination, and are required to keep a log of their hours using each gear type. On reaching 50 hours, the supervisor will sign and forward the log to the National Electro-fishing Safety and Training Administrator, who will issue a certificate for that gear type only.

For persons who have already qualified as Senior Operators with one gear type, a further 5 sessions will be the requirement to gain authorization for additional electro-fishing types. Trainees will need to keep a separate log of their hours of use with each gear type.

First-aid certificates and medicals must be updated every two years.

Safety Practices

A standard set of safety practices are listed below. Safety practices should include the following (adapted from H. G. Lazauski and Malvestuto, 1990).

1. All United States Coast Guard safety equipment for the operation of a 28 feet boat should be used.
2. Red Cross first aid and CRP training should be provided for all members of the electric fishing boat crew.
3. All members of the crew should be familiar with the electrical system of the boat.
4. All dip netters should wear rubber gloves, rubber boots, life vests and noise arresters if needed.
5. Boat operators should wear life vests, rubber boots and noise arresters if needed.
6. Electric fishing run should be kept under one hour to avoid netter's fatigue.
7. A strict check, through checklists, should be made of all electrical systems before each day's work in the field.
8. All fishing should cease at the first sign of lighting, rain, high winds or dip netter's fatigue.
9. Alcohol should never be allowed on an electric fishing boat.
10. Never touch the water or an electrode while the current is on.
11. Refuel the generator after engine is sufficiently cooled.
12. The boat driver should not make sudden turns or changes in boat speed.
13. No unauthorized passengers should ever be allowed on an electric fishing operation.
14. Know the range of your electric field. Avoid public recreation areas. Do not electro-fish near people or animals.

15. Avoid all unprofessional conduct (horse play).
16. Carry appropriate spare equipment for the boat.
17. Carry a first aid kit.

A detailed instruction guide or manual should accompany each electric fishing apparatus to assist the operator. The operator should be familiar with both the unit and manual before fishing begins. A log book should also be available to record dates and times of use, maintenance, problems and repairs.

An important emergency procedures is to have a pre-determined plan in the event of an accident. A documented route to medical facilities and procedures to follow is essential.

These safety procedures should be adhered to by all project personnel at all times. The safety check list and log book should be filled out every day. Also all operational parameters (control box settings and meter readings) should be recorded with field data and any observations of abnormal appearance, behavior, or mortality. This data will help refine parameters for future trips, avoid undesirable effects, and add to the data base on such effects. All members of the fishing crew should be familiar with the checklist material and compliance procedures.

Bibliography

Adams, W. J., Behmer, D. J. and W. O. Weingarten (1972) Recovery of shocked common shiner, Notropis cornutus related to electric energy, Transcations of the American Fisheries Society, 101(3);

Beaumont, W. R. C., A. A. L. Taylor, M. J. Lee and J. J. Welton (2002) Guidelines for electric fishing best practice, R and D Technical Report W2 - 054/TR

Benjamin, G. P. and C. C. Gillespie (1966) Type IV electro-fishing shocker, its characteristics and operation, SSP No. 529, Bureau of Commercial Fisheries, United States Dept. of Interior, Fish and Wildlife Service

Blasius and F. Schweitzer (1893) Beschreibung u. Benennung der "Galvano-narcose" Beobachtung der "Galvano-Hypnose" Pfluger's Archive 53: 527.

Biswas, K. P. (1970) Preliminary observations on the effect of electric on fish catch, Fishery Technology, Vol. VII, No.2

Biswas, K. P. (1971a) Determination of threshold current densities for different reactions of fishes using a pantostat, Fishery Technology, 8(1): 27 - 36.

Biswas, K. P. (1971b) Studies on threshold current densities for convulsion of fish using a square wave stimulator, Fishery Technology, 8(2): 143 - 155.

Biswas, K. P. and S. P. Karmarkar (1976) Reaction of fishes to the under water AC field Fishery Technology, 13(1): 26 - 35.

Biswas, K. P. (1996) Harvesting of Aquatic Resources, Daya Publishing House, Delhi – 110035, pp 1 - 199.

Biswas, K. P. (2015) Electrophysiology and fish behavior, Astral International Pvt. Ltd. New Delhi – 110002, pp 1 - 215.

Biswas, K. P. and N. A. Talwar (2017) On the development of an electric fish catcher Proceedings of the Zoological Soc,72(3), 246 - 256.

Biswas, K. P. (2019) Design and development of an electrically charged net for removing predators and unwanted fish from farm ponds, Fishing Chimes, 39(5), 46 - 54.

Biswas, K. P. (2019) Fishing with Electricity, Astral International Pvt. Ltd., New Delhi – 110002, pp 1 - 184.

Bjoringsoy, L. (1996) Gill netting – the selective fishing method, Infofish International 4: pp. 67 - 70.

Blanchteau, M, P. Lamarque, G. Mousset, and R. Vibert (1961) Etude neurophysioloique de la peche electrique. Extracted from Bull. Du Centre d'tudes et de recherché scientifiques, 3(3): 277 - 382.

Burnet, A. M. R. (1959) E#lectric fishing with pulsatory direct current, New Zealand Journal of Science, 2(1): 46 - 56.

Cain Jr, C. D. and J. W. Avault Jr (1983) Evaluation of a boat-mounted electro-trawl as a commercial harvesting system for crawfish, Aquacultural Engineering, 2(2): 135 - 152.

Chmielewiski, A., R. Cuinat, W. Dembinski and P. Lamarque (1973) Fatigue and mortality effects in electric fishing. Poiski Archiwun Hydrodiologil 20(2): 341.

Cowx, I. G. and P. Lamarque (1990) Fishing with electricity: application in freshwater fisheries management, Fishing News Books, Oxford (UK).

Denzer, H. W. (1949) Erfahrungen mit der Elektrofisherei, Fishereiwelt Jg 1: Hefts 3.

Denzer, H. W. (1956) Elektrofisherei, Stutgurt.

Dolan, C. R. and L. E. Miranda (2003) Immobilization thresholds of electro-fishing relative to fish size, Transactions of the American Fisheries Society, 132: 969 - 976.

Dolan, C. R. and L. E. Miranda (2004) Injury and mortality of warm water fishes immobilized by electro-fishing, North American Journal of Fisheries Management, 24: 118 - 127.

Emery, L. (1984) The physiological effects of electro-fishing, California-Nevada Wildlife Transactions,1984: 59 - 62.

Funk, J. L. (1947) Wider application of the electrical method of collecting fish, Transaction of the American Fisheries Society, Vol. 77.

Godfrey, H. (1956) Catches of fish in New Brunswick streams by direct current electro-fishing, The Canadian Fish Culturist, N. 19

Growns, I. O., D. A. Pollard and J. H. Harris (1996) A comparison of electric fishing and gill netting to examine the effects of anthropogenic disturbance on riverine fish communities, Fisheries Management and Ecology, 3: 13 - 24

Halsband, E (1959) The effect of pulsating direct current on fish, Modern Fishing Gear of the World pp. 575 - 580, Fishing News Books, London.

Halsband, E. (1954) Untersuchungen uber den Einfluss Verschiedener Stromarten auf den Stoffwechsel der Fische

Halsband, E. (1984) Electro-fishing (English translation of Halsband and Halsband, 1975 R. W. McCauley) Canadian Transaction in Fisheries and Aquatic Sciences, 5048.

Hartley, W. G. (1990) The history of electric fishing, Fishing with electricity, Applications in freshwater fisheries management, Fishing News Books, Blackwell Scientific Publications, Oxford, England, 1 - 3.

Hauck, F. R. (1949) Some harmful effects of the electric shocker on large rainbow trout Transactions of the American Fisheries Society, 77: 61 - 64.

Henry, T.B.,J. M. Grizzle, C. E. Johnson and J. A. Osborne (2004) Susceptibility of ten fish species to electroshock-induced mortality, Transactions of the American Fisheries Society, 133(3): 649 - 654.

Holzer, W. (1932) About an absolute voltage stimulus in fish, Physiology, 229; pp. 153 - 172, Berlin W.

Houston, R. B. Jr. (1949) Germen commercial electrical fishing device, US Dept of Int. Fish and Wildlife Service, Fish Leaflet, 348 pp 16.

Kuroki, T. (1953) Studies on the electric fish-screen VII. About the practical arrangements of electrodes to give proper distribution of potential Bulletin of the Japanese Society for the Science of Fish 18(9), pp 381 - 384.

Kuroki, T. (1959) Electrical Fishing in Japan, Modern fishing Gear of the World,pp 481 - 582, Fishing News (Books) Limited, London.

Lamarque, P. (1963) Electro-fishing: fish reactions with neurophysiological explanation National Science Progress (Paris), 3336: 137 - 148.

Loeb, J and S. S. Maxwell (1896), The theory of galvanotropism, Pfluger's Archive European Journal of Physiology, 63(3): 121 - 144.

Mac. Millan, F. O. (1928) Electric fish screen, Bulletin of the Bureau of Fisheries, 44(1042): 97 - 128.

Meyer-Waarden, P. F. (1955) Recent development in electro-fishing, F.A.O., World Fish Abstract, May/June, pp. 21.

Meyer - Waarden, P. F. (1957) Electrical fishing, 7(11): Food and Agriculture Organization of the United Nations.

Miyake, I and Steiger, W. R. (1957) The response of tuna and other fish to electrical stimuli, 223: U.S. Fish and Wildlife Service.

Myers, G. F. (1951) Design of an electric shocker boat, Progressive Fish Culturist, 13 (4): 229 - 231.

Pratt, V. S. (1951) A measure of the efficiency of an alternate and direct current fish shocker, Transaction American Fisheries Society, Vol. 81, 63 - 68.

Patten, B. G. and C. C. Gillespie (1966) Type IV Electro-fishing Shocker, The Bureau of Commercial Fisheries, (529), US Dept. of Interior.

Phillips, B.F. and Scolaro, A. B. (1980) An electro-fishing apparatus for sampling sublittoral benthic marine habitats, Journal of Experimental Marine Biology and Ecology, 47(1): 69 - 75

Price, A. L.and J. T. Peterson (2010) Estimation and modeling of electro-fishing capture efficiency for fishes in wadeable warm water streams, North American Journal of Fisheries Management, 30(2): 481 - 498.

Rosenberger, A. E. and J. B. Dunham (2005) Validation of abundance estimates from mark-recapture and removal techniques for rainbow trout captured by electro-fishing in small streams, North American Journal of Fisheries Management, 25(4): 1395 - 1410.

Ruetz, C. R., D. G. Uzarski, D. M. Kreuger and E. S. Rutherford (2007) Sampling a littoral fish assemblage; comparison of small mesh fyke netting and boat electro-fishing, North American Journal of Fisheries Management, 27(3): 825 - 831.

Scheminsky, F. (1922) Uber die Verschiedene Empfindlichkeit der Forelleneier wahrendihrer Entwicklung dem elektrischen strom gengenuber, Biochem. Ztschr. Bd.,132: S. 154 - 164.

Scheminsky, F. (1924) Experiments on electro-taxis and electro-anesthesia, Pfluger's Archive for the entire physiology of Man and Animals, 202(1); 200 - 216.

Schentiakov, V. A. (1963) Electrische Trawlversuche (Electric trawls), F.A.O. World Fish. Abstract 14(3): P 17.

Schreer, J. F., S. J. Cooke and K. B. Connors (2004) Electro-fishing - induced cardiac disturbance and injury in rainbow trout, Journal of fish Biology, 64(4): 996 -1014.

Sharber, N. G. and J. S. Black (1999) Epilepsy as a unifying principle in electro-fishing theory; a proposal, Transactions of the American Fisheries Society, 128: 666 -671.

Snyder, D. E. (2003) Invited overview conclusions from a review of electro-fishing and its harmful effects on fish; Fish Biology and Fisheries, 13: 445 - 453.

Smith, G. F. M and P. F. Elson (1950) A direct current electrical fishing apparatus, The Canadian Fish Culturists, No. 9: 1-13.

Stewart, P. A. (1975) Comparative fishing for Nephrops norvegicus (Linnaeus) using a beam trawl fitted with electric ticklers.

Tauti, M. (1931) A new form of electric fish screen, Journal of Imperial Fish Institution (Tokyo). 27(1): pp 33 - 44.

Vibert, R. (1963) Neurophysiology of electric fishing, Transactions of the American Fisheries Society, 92(3): 265 - 275.

Vibert, R. (1967) General report of the working party on the applications of electricity to inland fishery biology and management; Fishing with electricity. Its Application to Biology and Management, Fishing News (books), Limited, London, England, pp: 3 - 51.

Index